1993

# Siting Landfills and Other LULUs

## GEORGE NOBLE, P.E.

TECHNOMIC
PUBLISHING CO., INC.
LANCASTER · BASEL

## Siting Landfills and Other LULUs

a **TECHNOMIC**®publication

*Published in the Western Hemisphere by*
Technomic Publishing Company, Inc.
851 New Holland Avenue
Box 3535
Lancaster, Pennsylvania 17604 U.S.A.

*Distributed in the Rest of the World by*
Technomic Publishing AG

Printed in the United States of America
10  9  8  7  6  5  4  3  2  1

Main entry under title:
  Siting Landfills and Other LULUs

A Technomic Publishing Company book
Bibliography: p. 209
Includes index p. 211

Library of Congress Card No. 91-58006
ISBN No. 87762-878-5

### HOW TO ORDER THIS BOOK

BY PHONE: 800-233-9936 or 717-291-5609, 8AM-5PM Eastern Time
BY FAX: 717-295-4538
BY MAIL: Order Department
Technomic Publishing Company, Inc.
851 New Holland Avenue, Box 3535
Lancaster, PA 17604, U.S.A.
BY CREDIT CARD: American Express, VISA, MasterCard

To

Paula,

Tracy and David

# Contents

*Figures*    xi

*Tables*    xiii

*Preface*    xix

*Acknowledgements*    xxi

1. **INTRODUCTION** . . . . . . . . . . . . . . . . . . . . . . . . . . . . . . . . . . . . . . . . .1
   Traditional Landfill Siting    1
   Locally Unacceptable Land Uses    2

2. **SITING** . . . . . . . . . . . . . . . . . . . . . . . . . . . . . . . . . . . . . . . . . . . . . . .3
   Design with Nature    3
   Drastic    3
   Example    10
     *Depth to Water Table*  10
     *Net Recharge*  10
     *Aquifer Media*  11
     *Soil Media*  11
     *Topography*  11
     *Impact of Unsaturated Zone Media*  11
     *Hydraulic Conductivity*  11
   Pros and Cons    12
   Intrinsic Suitability    12
     *The First Six*  12
     *The Second Seven*  13
   Example    13
   Pros and Cons    14
   Review of Current Methods    14
   Sieving Techniques    15
     *The Mechanical Method*  15
     *Computerization*    15

3. **SITING CRITERIA** . . . . . . . . . . . . . . . . . . . . . . . . . . . . . . . . . . . . . .17
   Objectives    17
   Data Base    18

Methodology    18
Selecting the Criteria    18
Applying the Criteria    30
   *The Phased Approach*    30

4. **REGIONAL SITING CRITERIA** . . . . . . . . . . . . . . . . . . . . . . . . . . . . . . . . . . . . . . . . . . . . . . . . . **33**
Introduction    33
Natural Features    34
   *Wetlands and Waters of the United States*    34
   *Flood Plains*    36
   *Surface Waters*    37
   *Groundwater*    38
   *Suitable Soils for Groundwater Protection*    41
   *Fault Zones*    42
   *Seismic Impact Zones*    43
   *Unstable Areas*    46
   *Expansive Soils*    47
Land Use    49
   *Development—Existing & Committed*    49
   *Airports*    51
   *Municipal Wells*    55
   *Prime Farmland*    55
Economic Factors    56
   *Proximity to Major Highways*    56
Sieving    57
The Sieving Process    57
The Regional Siting Map    60
Sizing    62

5. **LOCAL SITING CRITERIA** . . . . . . . . . . . . . . . . . . . . . . . . . . . . . . . . . . . . . . . . . . . . . . . . . . **71**
Introduction    71
Natural Features    72
   *Depth of Suitable Soils for Cover*    72
   *Existing Depressions*    77
   *Natural Screening*    77
   *Run-On Potential*    78
   *Residential Well Density*    80
   *Ease of Monitoring Groundwater*    80
   *Slope*    80
   *Threatened and Endangered Species*    81
   *Scenic Areas*    81
   *Significant Depth to Groundwater Resources*    81
Land Use    82
   *Buffer Zone*    82
   *Final Use Compatibility*    82

*Municipal Boundaries* 83
*Area of Historic Importance* 83
*Areas of Architectural Importance* 84
*Areas of Paleontological Importance* 84
*Areas of Archaelogical Importance* 84
*Highway Restrictions* 84
*Traffic Impact* 86
*Distance from Centroid of Waste Generation* 86
*Availability* 86
*Land Holding in Large Parcels* 87
Local Siting      87
*Preliminaries* 87
Availability      92

6. **FIELD VERIFICATION** . . . . . . . . . . . . . . . . . . . . . . . . . . . . . . . . . . . . . . . . . **95**
Introduction      95
Criteria to Be Verified      95
Existing Depressions      95
Existing or Committed Development      95
Natural Screening      96
*Highway Users* 96
*Nearby Residents* 96
*Gathering Areas* 97
Run-On Potential      97
Threatened and Endangered Species      97
Highway Restrictions      98
Availability      98
Land Holding in Large Parcels      98

7. **WEIGHTING AND RATING** . . . . . . . . . . . . . . . . . . . . . . . . . . . . . . . . . . . . . **99**
Introduction      99
Weighting      99
*Depth of Suitable Soils for Cover* 101
*Existing Depressions* 101
*Natural Screening* 101
*Run-On Potential* 102
*Residential Well Density* 102
*Ease of Monitoring Groundwater* 102
*Scenic Areas* 103
*Significant Depth to Groundwater Resources* 103
*Buffer Zone* 103
*Final Use Compatibility* 104
*Municipal Boundaries* 104
*Highway Restrictions* 104
*Traffic Impact* 105

    *Distance from Center of Waste Generation or Transfer*   105
    *Availability*   105
    *Land Holding in Large Parcels*   105
  Rating   105
    *Depth of Suitable Soils for Cover*   106
    *Existing Depressions*   107
    *Natural Screening*   107
    *Run-On Potential*   108
    *Residential Well Density*   108
    *Ease of Monitoring Groundwater*   109
    *Scenic Areas*   110
    *Significant Depth to Groundwater Resources*   110
    *Buffer Zone*   111
    *Final Use Capability*   112
    *Municipal Boundaries*   112
    *Highway Restrictions*   113
    *Traffic Impact*   113
    *Distance from Center of Waste Generation*   114
    *Availability*   115
    *Land Holding in Large Parcels*   115
  The Question of Cost   116

**8. PUBLIC PARTICIPATION AND THE SITING PROCESS** . . . . . . . . . . . . . . . . . . . . . . . . . . . . **119**
  Introduction   119
  A Two Part Process   119
    *Regional Siting*   119
    *Local Siting*   120
  Problem Definition Meetings   120
  Regional Siting Meetings   122
  Local Siting Meetings   123
    *Shoe Leather Public Involvement*   123
    *Local Meetings*   125
  Public Participation Options   126
    *Option 1–The Siting Committee Only*   128
    *Option 2–Siting Committee + Siting Subcommittee*   128
    *Option 3–Siting Committee + Public Forum*   129
    *Option 4–Criteria Selection by Siting Committee/Weighting*
      *and Rating by Siting Committee + Siting Subcommittee*   129
    *Option 5–Criteria Selection by Siting Committee/Weighting*
      *and Rating by Siting Committee + Public Forum*   129
    *Option 6–Criteria Selection and Weighting by Siting*
      *Committee and Rating by a Rating Subcommittee*   129
    *Option 7–Criteria Selection and Weighting by Siting*
      *Committee/Rating by Public Forum*   129

Summary    130
After Weighting and Rating    130

## 9. CASE STUDY LAKE COUNTY, ILLINOIS . . . . . . . . . . . . . . . . . . . . . . . . . . . . . . . . . . . 131
Introduction    131
Introduction to the Lake County Solid Waste Management Plan    131
Natural Features    132
  *Floodplains*  132
  *Wetlands*  133
  *Threatened and Endangered Species*  133
  *Suitable Soils*  133
  *Surface Waters*  134
  *Surficial Aquifers*  134
Cultural/Institutional Resources    134
  *Historical Resources*  134
  *Archaeological/Paleontological*  134
  *Schools/Hospitals*  135
Development    135
  *Existing Development*  135
  *Committed Development*  135
Proximity to Major Highways    136
Additional Features    137
  *Airports*  137
  *Municipal Wells*  137
The Sieving Process    137
Composite Maps    139
  *Sanitary Landfill Composite*  139

## 10. SITING CRITERIA FOR OTHER LULUS . . . . . . . . . . . . . . . . . . . . . . . . . . . . . . . . . . . 145
Introduction    145
Transfer Stations    145
Incinerators    154
Compost Facilities    161
Sewage Treatment Facility    170
Recycling Facilities    176

## 11. THE WAY FORWARD . . . . . . . . . . . . . . . . . . . . . . . . . . . . . . . . . . . . . . . . . . . . . . . 185
The Environmental Coalition    185
Co-Opting Public Support    186
Building Public Trust    187
Contractual Agreements    188
  *Duration Limits*  188
  *Tonnage Limits*  190

Quality Assurance Programs
  *Liner Integrity*   191
  *Progress Reports*   191
Compensation Measures      192
  *Host Communities*   192
  *Private Citizens*   197
Advantages and Disadvantages      201
  *Mitigation*   201
  *Compensation*   202

Appendix A      205

References      209

Index      211

# Figures

| Figure 2-1 | A Digitizer for Use in Computer Mapping | 16 |
| Figure 4-1 | Measurement of Exclusion Distance From Runway | 52 |
| Figure 4-2 | Measurement of Exclusion Distances From Runway | 53 |
| Figure 4-3 | Criteria Mapping to Produce a Single Sieve Map | 58 |
| Figure 4-4 | Diagram Showing Positive and Negative Exclusion Zones | 59 |
| Figure 4-5 | A Typical Regional Siting Map | 61 |
| Figure 4-6 | Volume of the Frustrum of a Pyramid | 65 |
| Figure 4-7 | Landfill Geometry | 66 |
| Figure 4-8 | A Typical Regional Siting Map Showing How Small Areas are Excluded | 69 |
| Figure 5-1 | Soil Classification Chart | 73 |
| Figure 5-2 | Layered Landfill Cap | 75 |
| Figure 5-3 | Soils Suitable for Cover Material | 76 |
| Figure 5-4 | Typical Watershed Map | 79 |
| Figure 5-5 | Packer Trucks and Transfer Trailers Dimensions and Weight Limits | 85 |
| Figure 5-6 | A Typical Regional Siting Map | 88 |
| Figure 5-7 | A Typical Regional Siting Map Showing Preliminary Exclusion Zones | 90 |
| Figure 5-8 | Siting Map Showing Areas for Consideration in Local Siting Study | 91 |
| Figure 5-9 | Siting Map Showing Candidate Areas for Detailed Local Siting | 93 |

# FIGURES

Figure 9-1      Natural Features - Sieve 1      140

Figure 9-2      Cultural/Institutional Features - Sieve 2      141

Figure 9-3      Development - Sieve 3      142

Figure 9-4      Roads - Sieve 4      143

Figure 9-5      Sanitary Landfill Siting - Regional Study      144

# Tables

| Table 2-1 | Assigned Weights for Drastic Features | 4 |
|---|---|---|
| Table 2-2 | Ranges and Ratings for Depth to Water Table | 5 |
| Table 2-3 | Ranges and Ratings for Net Recharge | 5 |
| Table 2-4 | Ratings for Different Types of Aquifer Media | 6 |
| Table 2-5 | Ratings for Different Types of Soil Media | 7 |
| Table 2-6 | Ranges and Ratings for Topography | 7 |
| Table 2-7 | Ratings for Impact of Different Types of Unsaturated Zone Media | 8 |
| Table 2-8 | Ranges and Ratings for Hydraulic Conductivity | 9 |
| Table 2-9 | Computation of Best & Worst DRASTIC Scores | 10 |
| Table 3-1 | Operational Phases of Landfill Development | 19 |
| Table 3-2 | Environmental Impacts Associated with Landfill Development | 20 |
| Table 3-3 | Matrix of Operational Phases and the Environmental Impacts of Landfill Development | 21 |
| Table 3-4 | Potential Environmental Impacts of Landfill Development | 22 |
| Table 3-5 | Resources Sensitive to Landfill Development | 23 |
| Table 3-6 | Matrix of Environmental Impacts of Landfill Development and Potential Siting Criteria | 25 |
| Table 3-7 | Siting Criteria for Landfill Development | 29 |
| Table 3-8 | Regional Criteria for Landfill Siting | 31 |
| Table 3-9 | Local Criteria for Landfill Siting | 32 |
| Table 4-1 | Regional Criteria for Landfill Siting | 33 |

# TABLES

| | | |
|---|---|---|
| Table 4-2 | Criteria for Wetland Identification | 35 |
| Table 4-3 | Physical Properties of Prime Farmland | 55 |
| Table 4-4 | Compaction Densities | 62 |
| Table 5-1 | Local Criteria for Landfill Siting | 71 |
| Table 5-2 | Highway Restrictions of Concern for Solid Waste Haul Vehicles | 86 |
| Table 5-3 | Exclusion Criteria Included in Local Siting | 87 |
| Table 7-1 | Comparative Weights of Local Siting Criteria | 100 |
| Table 8-1 | Agenda for Stage 1 - Public Involvement | 121 |
| Table 8-2 | Agenda for Stage 2 - Public Involvement | 122 |
| Table 8-3 | Agenda for Stage 3 - Public Involvement | 126 |
| Table 8-4 | Public Participation Options | 128 |
| Table 9-1 | Landfill Siting Criteria | 138 |
| Table 10-1 | Construction and Operational Phases of Transfer Station Development | 146 |
| Table 10-2 | Environmental Impacts Associated With Transfer Station Development | 146 |
| Table 10-3 | Matrix of Construction and Operational Phases and Environmental Impacts for Transfer Stations | 147 |
| Table 10-4 | Potential Environmental Impacts of Transfer Station Development | 148 |
| Table 10-5 | Environments Sensitive to Transfer Station Development | 149 |
| Table 10-6 | Matrix of Environmental Impacts of Transfer Station Development and Potential Siting Criteria | 150 |

# TABLES

Table 10-7      Siting Criteria for Transfer Station Development      153

Table 10-8      Regional Siting Criteria for Transfer Station Development      153

Table 10-9      Local Siting Criteria for Transfer Station Development      154

Table 10-10     Construction and Operational Phases of
                Incinerator Development      154

Table 10-11     Environmental Impacts Associated With
                Incineration Development      155

Table 10-12     Matrix of Constructional and Operational Phases
                of Incineration Development Versus Environmental Impacts      156

Table 10-13     Potential Environmental Impacts of Incinerator
                Development      157

Table 10-14     Potentially Sensitive Environments and Other Factors
                Limiting Incineration Development      158

Table 10-15     Matrix of Environmental Impacts of Incineration and
                Potential Siting Criteria      159

Table 10-16     Selection Criteria for Incineration Development      160

Table 10-17     Regional Siting Criteria for Incineration Development      160

Table 10-18     Local Siting Criteria for Incineration Development      161

Table 10-19     Construction and Operational Phases of Compost
                Facility Development      161

Table 10-20     Environmental Impacts Associated With Compost
                Facility Development      162

Table 10-21     Matrix of Construction and Operational Phases
                and the Environmental Impacts of Compost Facilities      163

Table 10-22     Potential Environmental Impacts of Compost Facility
                Development      164

| Table 10-23 | Environments Sensitive to Compost Facility Development | 165 |
| Table 10-24 | Matrix of Environmental Impacts of Compost Facility Development and Potential Siting Criteria | 166 |
| Table 10-25 | Siting Criteria for Compost Facilities | 168 |
| Table 10-26 | Regional Criteria for Compost Facility Siting | 169 |
| Table 10-27 | Local Criteria for Compost Facility Siting | 169 |
| Table 10-28 | Construction and Operational Phases of Sewage Treatment Facility | 170 |
| Table 10-29 | Environmental Impacts Associated With Sewage Treatment Facility Development | 170 |
| Table 10-30 | Matrix of Construction and Operational Phases of Sewage Treatment Facilities Versus Environmental Impacts | 171 |
| Table 10-31 | Potential Environmental Impacts of Sewage Treatment Facility Development | 172 |
| Table 10-32 | Potentially Sensitive Environments and Other Factors Limiting Sewage Treatment Facility Development | 173 |
| Table 10-33 | Matrix of Environmental Impacts of Sewage Treatment Facility and Potential Siting Criteria | 174 |
| Table 10-34 | Selection Criteria for Sewage Treatment Facility Development | 175 |
| Table 10-35 | Regional Siting Criteria for Sewage Treatment Facility Development | 175 |
| Table 10-36 | Local Siting Criteria for Sewage Treatment Facility Development | 176 |
| Table 10-37 | Construction and Operational Phases of Recycling Facility Development | 176 |

# TABLES

Table 10-38    Environmental Impacts Associated With Recycling
Facility Development                                           177

Table 10-39    Matrix of Construction and Operational Phases of Recycling
Facility Development Versus Environmental Impacts              178

Table 10-40    Potential Environmental Impacts of Recycle Facility
Development                                                    179

Table 10-41    Potentially Sensitive Environments and Other Factors
Limiting Recycling Facility Development                        180

Table 10-42    Matrix of Environmental Impacts of Recycling Facilities
and Potential Siting Criteria                                  181

Table 10-43    Selection Criteria for Recycling Facility Development      182

Table 10-44    Regional Siting Criteria for Recycling Facility Development 182

Table 10-45    Local Siting Criteria for Recycling Facility Development   183

Table 11-1     Items of Landfill Development Necessary for Operation
Regardless of Size and Duration                               189

Table 11-2     Items of Landfill Development Directly Related to
Landfill Size                                                 190

# Preface

It is a sign of the times we live in that many people will recognize the acronym LULU as signifying Locally Unacceptable Land Uses. The acronym seems to be appearing everywhere while the LULU itself is, of course, rarely appearing anywhere. Resistance to these various necessary features of modern development is now so pervasive, that few LULUs are sited without widespread controversy, and a great many are never sited at all.

As far as the siting procedure is concerned, the battle rages over stealth versus open discussion. Some appear to favor deals behind the scenes, followed by a shock announcement. This approach is generally designed to mask the fact that there has been little or no real site selection process. The opposing camp favors all out public involvement. This approach generally indicates at least a reasonable level of planning.

The whole thesis of this book is that with careful planning, it _is_ possible to get LULUs sited. My experience in this area has lead me over the years to the conclusion that it is vital to trust the public in this area and include them fully in the siting process. It is axiomatic that the public will be involved sooner or later. That they should be involved early may be harder for some landfill engineers to swallow, but I hope to show in these pages that it is essential to ultimate success. It is also absolutely crucial that homework on the subject of siting should be completed fully and thoroughly, because once invited into the process, the public can be a hard taskmaster.

What is described in these pages is essentially organized homework. The process requires development of siting criteria from first principles. If the process is followed completely, the landfill engineer will cover all the aspects of siting necessary to be able to manage all levels of intensive public involvement.

The process also involves complete recognition that the public has a substantial case for skepticism when presented with reassurances of safety. Practical mitigation measures are presented in order to allay these public fears. If the landfill engineer is certain there are no risks, then there should be nothing to fear from establishing full disclosure reporting procedures, and penalties in the event that problems do occur.

Wilmette, Illinois
May 1991

# Acknowledgements

Several professional colleagues have contributed significantly to this book by their influence and steady judgement over the years. In particular I wish to thank Richard Eldredge, Keith Gordon, and James Tracy. I acknowledge their guidance, and absolve them from responsibility for any mistakes which may have escaped the review process.

To the members of the Lake County Solid Waste Action Agency, I owe my thanks for a timely contract to perform regional siting in Lake County. This contract helped to get the book moving again when it seemed that it was irretrievably stuck somewhere between intention and too little time. I am particularly indebted to Bill Barron, the Assistant Administrator of Lake County, for his unflappable good nature in handling the intricacies of the Lake County project with grace, and a steady eye for the overall goals of the project. I am indebted to the Agency for their permission to use the regional siting maps generated during that study as examples of how this siting system works in practice. I stress that the maps used in Chapter 9 from the Lake County study are the results of a regional study only, and local siting criteria for Lake County are not presented here.

I wish to thank my secretary, Penny Klopp, for painstakingly preparing draft after draft of this book for editing and re-editing, and for doing so with her usual thoroughness and good humor.

To my daughter, Tracy, I wish to say thank you for a devastatingly thorough review of an early draft which helped considerably in preparation of the final version. Any lack of clarity remaining is entirely due to my own shortcomings in spite of her best intentions to correct the problems. I also wish to thank Tracy for all her work on producing the matrices in Chapter 3 and Chapter 10.

Finally, I wish to say a particular thank you to my wife, Paula, for bearing with the many early mornings, late nights, and weekends spent in working on the book. I appreciate her encouragement and support for the whole project.

# Introduction

## TRADITIONAL LANDFILL SITING

The tradition of using marginal land for landfill dies hard. Land with little or no economic value was for many years considered good for nothing but garbage disposal. However, the specific characteristics which have lead in the past to the rejection of marginal land for other uses render it utterly useless for garbage disposal. Time and again we find that the reason for landfill failure and subsequent widespread contamination has its roots in the original decision to choose a bad site without the natural protection which would have at least partially compensated for poor design.

Marginal land is not good enough for garbage disposal.

Marginal land is land deficient in some natural feature which renders it of little value for a wide range of uses. For example, ravine land is marginal land because the geology or natural erosion have created an area of steep topographic drainage. Such an area would be impossible to work for agricultural uses without a great deal of earth moving, and impossible to build upon without significant expenditures in extra support. Landfills constructed in ravines were common twenty years ago, and we are still living with the consequences of rapid leachate movement and often slippage of the entire fill as it travels down the path of natural drainage.

Quarried land is marginal land because when sand, gravel or limestone have been removed the resultant excavation requires massive filling before any normal use. Landfills constructed in worked-out quarries are common even today, and yet they are often sited over important sources of water supply; areas which would be avoided with proper siting.

Marginal land is often unused and unwanted and often sits idle for many years without human intervention. This neglect is attractive to one use generally forgotten in the planning process and that is wildlife. Marginal sites are often colonized by wildlife species increasingly threatened with extinction because of dwindling habitat. It is ironic that such sites are thereby protected by receiving designation as habitat for threatened and endangered species.

However marginal land is not the only type of land which is not good enough for landfill. In fact many types of land which are good for a wide variety of uses are simply not capable of providing the kind of environmental protection needed for landfill. Modern landfill is an extremely specialized use for land and the selection of a site should be conducted with as much care and attention to detail as the design.

# LOCALLY UNACCEPTABLE LAND USES

There are in fact many other special uses for land which require the same care in site selection. In addition to landfills these special uses include:

> Transfer Stations
> Incinerators
> Compost Facilities
> Sewage Treatment Facilities
> Recycling Facilities

The list goes on. All of these special uses have one thing in common and that is that they are considered by many people to be Locally Unacceptable Land Uses or LULUs. The term is of recent origin, although the phenomenon has been around for centuries.

The reasons these land uses are considered unacceptable are many and various, but not least is the fact that past experiences have generally indicated that they cause significant environmental problems. All too often, however, the problems are caused by bad siting, and not some unacceptable feature inherent in the special use itself.

Proper siting is vital if these special uses are to be turned into Locally Acceptable Land Uses Properly Sited.

Increasingly LULUs are required to go through extended public review before they are approved. Often proper siting is a major element of the review process. Without proper siting, a design may never have the chance to prove itself no matter how resourceful it may be. It makes no sense to expend considerable time and energy to design an innovative project if the absence of a proper siting process will kill it.

This book seeks to take the pain out of the siting process, and establishes the first principles of siting so that they can be applied to any difficult-to-site project.

# Siting

## DESIGN WITH NATURE

From the early days of the town planning movement the ideal resolution of land use problems has been to identify the highest and best use for every piece of land that comes under the sway of the planners rule.

Early planners such as Francis Law Olmsted, Sir Patrick Geddes, and Ebenezer Howard sought to identify the distinguishing fingerprint of an area of real estate which marked it for a special use. Ian McHarg in his book *Planning With Nature* showed how this might be achieved by mapping the various aspects of large metropolitan areas on huge acetate sheets, and overlaying them to sieve-out the salient features which point the way to future plans.

Clearly when applied to metropolitan areas the process is complex and frustrating because there is so much which the planner must accept as a given. When applied to the selection of a site for a LULU however, the process is relatively straight forward. Here the exercise is to find the areas which development has overlooked and choose the best candidate for use as a landfill, incinerator or whichever LULU is under consideration.

This book takes as it's starting point the landfill siting problem, and seeks to show how this process works in some detail. Later the book extends the same procedure and points the way to the application of the same procedure for other LULUs.

The landfill siting problem has been around for a long time. How has it been handled in the past? What can we learn from these methodologies?

We know from almost daily stories of massive opposition to landfill siting that the sure way to failure is to have no siting system.

We know that there are also many other siting systems aside from the one espoused in later chapters of this book. Some of these systems will be discussed here as there is much to learn from their general approach.

## DRASTIC

A criteria selection system called DRASTIC has been developed by the U.S. Environmental Protection Agency (EPA) and the National Water Well Association (EPA/NWWA 1985) for evaluating groundwater pollution potential using hydrogeologic settings. A wide range of siting criteria were considered in the development of this scheme, however, the availability of mapping data was a major concern, and in the final system only criteria which are readily available in map form were used. The system compares areas by assigning ratings and weights to seven parameters that affect groundwater contamination. Briefly, these seven parameters are:

D - Depth to Water Table
R - Recharge (net infiltration)
A - Aquifer Media
S - Soil Media (surface soils)
T - Topography (Slope)
I - Impact of the Unsaturated Zone Media
C - Conductivity (Hydraulic) of the Aquifer

These factors have been arranged to form the acronym, DRASTIC for ease of reference.

Each DRASTIC factor has been evaluated with respect to the other to determine the relative importance of each factor. Each DRASTIC factor has been assigned a relative weight ranging from 1 to 5 (Table 2-1). The most significant factors have weights of 5; the least significant, a weight of 1.

The depth to water and the impact of the unsaturated zone media are considered to be the most important criteria in the seven. These two criteria are followed in order of importance by net recharge.

Each DRASTIC factor has been divided into either ranges or significant media types which have an impact on pollution potential. Each of these ranges or media types is then assigned a rating which varies between 1 and 10. See Tables 2-1 to 2-8. The highest rating indicates the highest pollution potential. For example, in Table 2-2 if the depth to Water Table is in the range 0-5 feet, the pollution potential rating is 10. On the other hand, if there is over 100 feet to the water table, the pollution potential is only 1.

TABLE 2-1
ASSIGNED WEIGHTS FOR DRASTIC FEATURES

| FEATURE | WEIGHT |
|---|---|
| Depth to Water Table | 5 |
| Recharge | 4 |
| Aquifer Media | 3 |
| Soil Media | 2 |
| Topography | 1 |
| Impact of the Unsaturated Zone Media | 5 |
| Hydraulic Conductivity of the Aquifer | 3 |

4

## TABLE 2-2
## RANGES AND RATINGS FOR DEPTH TO WATER TABLE

### DEPTH TO WATER

| RANGE IN FEET | RATING |
|---|---|
| 0-5 | 10 |
| 5-15 | 9 |
| 15-30 | 7 |
| 30-50 | 5 |
| 50-75 | 3 |
| 75-100 | 2 |
| 100+ | 1 |

## TABLE 2-3
## RANGES AND RATINGS FOR NET RECHARGE

### NET RECHARGE

| RANGE IN INCHES | RATING |
|---|---|
| 0-2 | 1 |
| 2-4 | 3 |
| 4-7 | 6 |
| 7-10 | 8 |
| 10+ | 9 |

## TABLE 2-4
## RATINGS FOR DIFFERENT TYPES OF AQUIFER MEDIA

| MEDIUM | RATING | TYPICAL RATING |
|---|---|---|
| Massive Shale | 1-3 | 2 |
| Metamorphic/Igneous | 2-5 | 3 |
| Weathered Metamorphic/Igneous | 3-5 | 4 |
| Glacial Till | 4-6 | 5 |
| Bedded Sandstone, Limestone and Shale Sequences | 5-9 | 6 |
| Massive Sandstone | 4-9 | 6 |
| Massive Limestone | 4-9 | 6 |
| Sand and Gravel | 6-9 | 8 |
| Basalt | 2-10 | 9 |
| Karst Limestone | 9-10 | 10 |

TABLE 2-5
RATINGS FOR DIFFERENT TYPES OF SOIL MEDIA

| MEDIUM | RATING |
|---|---|
| Thin or Absent | 10 |
| Gravel | 10 |
| Sand | 9 |
| Peat | 8 |
| Shrinking and/or Aggregated Clay | 7 |
| Sandy Loam | 6 |
| Loam | 5 |
| Silty Loam | 4 |
| Clay Loam | 3 |
| Muck | 2 |
| Nonshrinking and Nonaggregated Clay | 1 |

TABLE 2-6
RANGES AND RATINGS FOR TOPOGRAPHY

| TOPOGRAPHIC RANGE IN PERCENT SLOPE | RATING |
|---|---|
| 0-2 | 10 |
| 2-6 | 9 |
| 6-12 | 5 |
| 12-18 | 3 |
| 18+ | 1 |

7

## TABLE 2-7
## RATINGS FOR IMPACT OF DIFFERENT TYPES
## OF UNSATURATED ZONE MEDIA

| MEDIUM | RATING | TYPICAL RATING |
|---|---|---|
| Confining Layer | 1 | 1 |
| Silt/Clay | 1-2 | 3 |
| Shale | 2-5 | 3 |
| Limestone | 2-7 | 6 |
| Sandstone | 4-8 | 6 |
| Bedded Limestone, Sandstone, Shale | 4-8 | 6 |
| Sand and Gravel with Significant Silt and Clay | 4-8 | 6 |
| Metamorphic/Igneous | 2-8 | 4 |
| Sand and Gravel | 6-9 | 8 |
| Basalt | 2-10 | 9 |
| Karst Limestone | 8-10 | 10 |

TABLE 2-8
## RANGES AND RATINGS FOR HYDRAULIC CONDUCTIVITY

| HYDRAULIC CONDUCTIVITY RANGE IN GPD/FT $^2$ | RATING |
|---|---|
| 1-100 | 1 |
| 100-300 | 2 |
| 300-700 | 4 |
| 700-1000 | 6 |
| 1000-2000 | 8 |
| 2000+ | 10 |

This system allows the user to determine a numerical value for any hydrogeologic setting by using an additive model. The equation for determining the DRASTIC Index is:

$$D_R D_W + R_R R_W + A_R A_W + S_R S_W + T_R T_W + I_R I_W + C_R C_W = \text{Pollution Potential}$$

Where the subscripts R and W indicate the rating and the weight respectively as identified for each quantity in the drastic index.

In Summary:

R = rating
W = weight

To return to the example derived from Table 2.2, we already know that Depth to Water carries a weight of 5. If the depth to water table is in the range 0-5, then the rating is 10. Thus, in this example, the total contribution towards the pollution potential is:

5 x 10 = 50

DRASTIC is basically a comparative tool which is used to compare one site to another. However, the system can also be utilized to compare a given site with idealized scores in order to judge comparatively whether or not a site has merit.

## TABLE 2-9
## COMPUTATION OF BEST AND WORST DRASTIC SCORES

| DRASTIC | WEIGHT | RATINGS BEST | WORST | RATINGS BEST | WORST |
|---|---|---|---|---|---|
| Depth of Water Table | 5 | 1 | 10 | 5 | 50 |
| Recharge | 4 | 1 | 9 | 4 | 36 |
| Aquifer Media | 3 | 2 | 10 | 6 | 30 |
| Soil Media | 2 | 1 | 10 | 2 | 20 |
| Topography | 1 | 1 | 10 | 1 | 10 |
| Impact of the Unsaturated Zone Media | 5 | 1 | 10 | 5 | 50 |
| Hydraulic Conductivity | 3 | 1 | 10 | 3 | 30 |
| TOTAL | | | | 26 | 226 |

The computation of best and worst scores possible is identified in Table 2-9.

EXAMPLE

A criterion-by-criterion discussion of the selected ratings for a hypothetical landfill site follows. Consider the case of "Hellova Landfill".

Depth to Water Table

According to data available the depth to water table is 15-40 feet below the original land surface.
Rating = 6

Total Score = 5 x 6 = 30

Net Recharge (weight = 4)

According to a design report let us say that the area experiences an infiltration rate of 70% of annual precipitation and that the net recharge works out to be greater than 10 inches. According to Table 2-2, a net recharge of greater than 10 inches has a rating of 9.

Total score = 4 x 9 = 36

10

<u>Aquifer Media</u> (weight = 3)

The site consists of fractured carbonate bedrock aquifer overlain by sand and gravel.
Rating = 8

Total score = 3 x 8 = 24

<u>Soil Media</u> (weight = 2)

There is sand at the surface.
Rating = 8

Total score = 2 x 8 = 16

<u>Topography</u> (weight = 1)

6-12% is a conservative estimate of the range of slopes on the site.
Rating = 5

Total score = 1 x 5 = 5

<u>Impact of Unsaturated Zone Media</u> (weight = 5)

Sand and gravel with some silt and clay.
Rating = 7

Total score = 5 x 7 = 35

<u>Hydraulic Conductivity</u> (weight = 3)

Laboratory and single well tests suggest field values in the $10^{-4}$ to $10^{-3}$ range for the sand units.

Rating = 8

Total score = 3 x 8 = 24

The total DRASTIC score for the Hellova Landfill is: 170.

The best score is 26.

The worst score is 226.

In the range of 26 to 226, a score of 170 is 72% above the best possible score. In any regionwide screening process, this site would not be expected to win approval.

## PROS AND CONS

The advantage of this method is that maps are readily available for each of the chosen criteria, so with relatively few resources a community could apply this DRASTIC Index and compare several sites.

The disadvantage is that the procedure is limited to only seven criteria and there are so many more to be considered that the results of a DRASTIC evaluation could only be considered a small step in the process. The second and more significant disadvantage is that these seven criteria are not the most important criteria. There are, for example, several criteria which are so important that they are often considered to be exclusionary, that is to say, if any one of these factors exists at the site then the site should be excluded from any further consideration.

Take for example the question of fault zones. It is commonly recommended that landfills should not be located within 200 feet of a fault zone that has had displacement in Holocene time (roughly in the last 11,000 years). This criterion is not a function of DRASTIC. One can imagine the reaction in a local community which has spent several months of anguish in comparing sites by the DRASTIC method only to find that the highest rated site has to be discarded because it happens to lie over a major fault zone.

This criticism raises a third objection and that is the weighting process itself. Stated simply if an area is excluded by reason of one criterion why go through the trouble of judging what weight to give it.

## INTRINSIC SUITABILITY

Before 1989 the Minnesota Pollution Control Agency (MPCA) had a system which attempted to solve the problem of exclusionary and non exclusionary factors. In Minnesota there were thirteen criteria which had been adopted by the MPCA to evaluate suitability for sanitary landfill. The criteria fell into two distinct categories. The first six criteria were exclusionary and failure to meet any one of these six resulted in exclusion from further consideration. The next seven criteria could be overcome by engineering. The important consideration in administering each of these criteria was: how well does the engineering overcome the deficiency? The criteria were as follows:

### The First Six

The fill and trench areas of the proposed landfill area cannot be within:

   (1)   1,000 feet of the normal high water mark of a lake, pond or flowage;

   (2)   300 feet of a stream;

   (3)   a regional (100 year) floodplain; or

   (4)   a wetland.

In addition to the four (4) criteria mentioned above, a site is not intrinsically suitable if:

(5)   it would present a bird hazard to an airport; or

(6)   there is karst development on the site.

## The Second Seven

These seven (7) secondary criteria can also render a site intrinsically unsuitable depending upon the extent to which engineering measures can be relied upon to overcome them. These seven criteria are:

(1)   proposed fill and trench areas are within 1000 feet of the nearest edge of the right-of-way of any state, federal, or interstate highway; any public park or an occupied dwelling;

(2)   any wetlands or public waters would be impacted during development of the site;

(3)   there are erosion, drainage or other natural processes occurring in the area which could lead to problems at the site or site failure;

(4)   a drinking water supply reservoir would be impacted by the site;

(5)   any ground water which is present is:

(a)  a water supply
(b)  is capable of being withdrawn at a sustained yield of one gallon per minute
(c)  recharging to another aquifer;

(6)   ground water is not protected by an aquiclude; and

(7)   ground water cannot be monitored by routine methods.

## EXAMPLE

Consider the case of Hellova Landfill. Let us assume that this site is not excluded by any of the first six criteria. Having passed the first test we move on to the second seven criteria.

(1)   The site is 500 feet from the highway, but the proponent proposes a high berm with fast growing hybrid trees at the top to limit the overview.

(2)   There are no wetlands or public waters on the site.

(3)   There are no erosion, drainage, or other natural processes in the area which could lead to problems.

13

(4)   There is no drinking water supply reservoir which would be impacted by the site.

(5)   The area is a natural recharge zone for an existing municipal water supply. The possibility exists of importing clay to the site and recompacting it to form an artificial aquitard, but this is not considered adequate protection and the site is excluded from further consideration.

## PROS AND CONS

The advantage of this system is that it covers many of the exclusionary criteria and applies them first. Thus, the ability to narrow the field of study is built into this approach.

There are several disadvantages. The first concerns item one of the second seven. The exclusion of sites within 1000 feet of any state, federal or interstate highway seems to be unduly broad even though it does allow for mitigation.

In many siting studies actual proximity to state, federal or interstate highways is a positive criterion because it minimizes the cost of construction of the access road. In almost all cases proximity to a highway can be mitigated by a combination of berms and landscaping treatment such that this element of criterion (1) is easily overcome.

A second disadvantage is that there is no exclusion for valuable wildlife habitat.

A third disadvantage is that the second seven all depend upon an exercise of judgement and without some systematic weighting process the application of the second seven can fuzzy and contentious.

## REVIEW OF CURRENT METHODS

Although there are many different lists of criteria, the two presented above are indicative of the selection procedures available. In general, there are clearly some criteria which are so important that they preclude landfill no matter what mitigation is considered. These criteria should be mapped as exclusion zones. There are many other criteria which are simply problematic and these must be considered by application of a weighting process which clarifies their level of importance compared to one another.

From these concerns grows the two part application of siting criteria espoused in this book. The first part applies the exclusionary items in order to narrow the area of search. This is often referred to as the Regional Study. The second part applies the remaining criteria in an effort to locate specific sites. This is the time for weighting since none of these second round factors is necessarily exclusionary. This round is often referred to as the Local Study.

# SIEVING TECHNIQUES

## The Mechanical Method

As mentioned earlier the early planners used large acetate overlays to sieve out the desired criteria. There is no reason why this technique should not continue to be used, particularly for small areas, although it can be extremely tedious when applied to large areas. Take for example the first of the above criteria: Depth to Water Table.

Many areas of the country have aquifer maps available which identify depth to water table and this information can very simply be applied to an acetate sheet by using a wax pencil or by transferring the information photographically to a sheet of clear overlay at the required scale.

Obviously, some attention needs to be paid to choosing a scale. The maps need to be manageable as tools for choosing prospective sites. The maps also need to be capable of use at meetings and hearings to illustrate how areas were selected. They must also be capable of reduction into the form of a report.

Here we begin to see some of the practical difficulties of this technique because in practice the scale needs to be large enough to identify local features such as roads, bridges, houses and other aspects of land use while capable of easy reduction to be comprehensible in a report.

When you add an additional criterion - flexibility, the acetate overlay process becomes very unwieldy.

The system needs to be flexible because it may be necessary to change criteria at some point. It needs to be flexible because new information may alter the map.

## Computerization

Since we first began to receive geographical information from EARTHSAT the computer has been increasingly used as a vehicle for storage of geographical data. While the initial work of transferring mapped information to the computer is very labor intensive, the ability to vary scale and select specific data more than repays the initial investment.

Geographical information is stored by the use of a digitizer which is a small magnifying glass with a painted target at the center. The digitizer also contains several buttons, one of which is pressed by the operator when specific information is to be recorded. See Figure 2-1. The digitizer can only be used on a digitizing board which is a large drawing board implanted with an electronic array capable of identifying precisely the location of the digitizer on its surface.

The outline of a specific item of geographical information is rendered as an irregular polygon by the digitizer rather in the way children connect dots to make a picture. A computer record is made at each dot. Each dot represents that point where the straight line changes direction. Under magnification the work of a digitizer looks like a series of straight lines connecting dots. A conscientious digitizer can render a map to a very close approximation to the original.

Once the work of the digitizer is completed it is a simple matter to "overlay" the criteria maps and produce a composite or sieve map which shows only those areas left over after all excluded areas have been omitted.

FIGURE 2-1

A DIGITIZER FOR USE IN COMPUTER MAPPING

PHOTO BY PERMISSION OF INTERGRAPH CORPORATION

# 3

# Siting Criteria

## OBJECTIVES

In selecting criteria for site selection, it is important to identify the objectives of the site selection process. One inadmissible objective is to stop landfill siting. It is entirely inappropriate to use the site selection process to show that there are no suitable sites. In fact, as we shall see, within certain constraints, the criteria may be adjusted, consistent with changes in landfill design, to increase the number of sites to be considered. Although it may appear entirely obvious in siting a landfill, only one objective is assumed and that is:

> A real need to site

As we will see later, this assumption has significant impacts upon the methodology throughout the siting process. A real need to site means that a "zero site option" is not realistic. A real need to site has important implications to the selection of the size of the region under study. Clearly, a suburban community with no undeveloped land, bounded on all sides by other similar communities, is wasting time and money identifying its own community as the area within which a site will be selected. What is required in circumstances of this kind is an inter-agency compact large enough to offer some reasonable possibility of finding undeveloped land suitable for use as a landfill. *The purpose of the siting process is to make the best use of the land resources available.*

If a zero site option is unrealistic, this also means that the siting body must be willing to alter the selection criteria in order to come up with real site options.

If the criteria can be altered so readily to find a site, what is so important about the process? Why even bother? Fair questions if landfills or other LULUs were rigidly defined and unalterable in their impact upon the environment. However, the fact is that the design of these facilities can be extremely flexible. For example if the application of the landfill siting criterion which identifies areas of clay greater than 50 feet in thickness reveals no areas meeting this criterion, the siting criteria should be altered to 40 feet, or whatever thickness (within reasonable limits) in order to identify realistic options. The loss of this additional degree of protection then requires the redefinition of the landfill design criteria. In this example, the minimum design for the landfill might choose to compensate for the loss of an additional 10 feet of clay by recompacting the top 10 feet to a lower permeability in order to provide the same net protection as 50 feet of natural clay.

Alternatively, the minimum design for the landfill might compensate for this loss by the addition of a flexible membrane liner to make up for the reduced thickness of clay. In certain states the existing clay may be excavated and recompacted to improve the permeability, or additional clay may be imported to compensate for the reduced clay separation. The ability to modify the design of the landfill to meet the demands of the region under study is a very important aspect of the siting process. In a later chapter, possible modifications to the landfill design will be suggested for each of the criteria identified in those cases where the criteria have to be modified.

The objective of having a real need to site has the effect of modifying the criteria in many ways. Basically, the criteria must be flexible in order to develop siting options, and the "price" of each modification must be identified with changes in the landfill design criteria.

## DATA BASE

The criteria selected will also be influenced by the data base. Stated simply, use whatever good data is available that is capable of identifying broad general areas of exclusion.

For example, in many states, groundwater data is extremely limited. The absence of this data should not derail the site selection study. As we will see later, the selection process is conducted in two parts; regional, then local. Whatever criteria are excluded from the regional study, due to deficiencies in the data base, should be included in the local study. A regional siting study merely serves to limit the areas which need to be studied in greater detail later.

Certain criteria are favorable to landfill siting, while others are unfavorable. For the purposes of the mapping process itself, the data being mapped is essentially the same whether the criterion is favorable or unfavorable since what is unfavorable is merely the area remaining after mapping the favorable. For example, it is generally favorable for landfill siting to search for areas where soils exist with permeability of $1 \times 10^{-7}$ cm/sec and thickness greater than 50 feet. This criterion is so commonly accepted as a defense against groundwater contamination that many state geological survey agencies have such maps readily available. For _mapping_ purposes we map all areas which have permeability of $1 \times 10^{-7}$ cm/sec, and the thickness greater than 50 feet. For the purposes of map _interpretation_, we are only interested in excluding all areas which do _not_ have permeability of $1 \times 10^{-7}$ cm/sec and thickness less than 50 feet.

In some cases the exclusion areas are mapped directly. For example, wetland areas, and areas within 200 feet of wetland areas, are unfavorable areas for landfill siting, and are directly mapped as exclusion zones, or areas to be excluded from further study.

## METHODOLOGY

The method to be used to map data (discussed earlier in Chapter 2) will also have an impact on the criteria selection. Computer mapping is extremely flexible, and the choice of criteria can be made on a "rough cut" basis initially, since a later change can be easily accomplished.

Changes in sieve maps prepared by hand are much more difficult. Thus, the choice of siting criteria must be reasonably certain before starting the mapping process if they are to be prepared by hand.

## SELECTING THE CRITERIA

Site selection criteria for landfills, or any other LULUs are essentially mirror images of impact mitigation factors. In other words, the site selection criteria should be chosen to minimize or eliminate the negative impacts associated with landfills. Site selection criteria can, therefore, be developed in the same way that environmental impact statements are checked for the completeness of coverage. The method requires the building of a two stage matrix.

The first is a matrix of Operations vs Impacts. This matrix helps to determine specifically what items of operation cause specific impacts. To many looking at this for the first time, the determination of what operation causes what impact may be numbingly obvious. I can only say, try it conscientiously, and you will be surprised what a powerful organizing tool matrix building can be.

The second matrix is more difficult because it involves building a general list of land use factors which may be sensitive to the environmental impacts. In a general text such as this, the list of sensitive land use factors is necessarily broad. In a specific region of the country, this list may not need to be as exhaustive. However, even in building this second matrix, make an effort to extend the list to be as specific as possible. Again, if you try this matrix building conscientiously, I am sure you will find it rewarding. Those who do not care for the matrix method have the option of simply using the criteria lists presented here and performing the necessary mapping.

A possible middle course might be to simply start with the criteria lists, and modify them to be sure they meet all local conditions.

We may start the process of criteria selection by making a list of those factors in the operation of a landfill which generate known or even perceived impacts as shown in Table 3-1.

Next review these operational phases, and identify the attendant environmental impacts. If the impact factors suggest a new sub category of operation, return to Table 3-1, and make the change. In turn this may suggest changes in the list of environmental impacts which is shown in Table 3-2.

TABLE 3-1
OPERATIONAL PHASES OF LANDFILL DEVELOPMENT

Haul to Landfill
Queue at Landfill
Excavation
Garbage Deposition
Compaction
Soil Cover
Landscaping
Leachate Generation
Leachate Treatment
Gas Generation

TABLE 3-2

---

ENVIRONMENTAL IMPACTS ASSOCIATED WITH LANDFILL DEVELOPMENT

Use Of Land
Highway Congestion
Highway Noise
Accidents
Noise
Air Pollution
Rats
Flies
Groundwater Contamination
Economic Loss
Reduction of Scenic Value

---

These two lists mounted together form a matrix as shown in Table 3-3. Review each phase of the operation, and mark those items which will give rise to specific environmental impacts.

## TABLE 3-3:
## MATRIX OF OPERATIONAL PHASES AND THE ENVIRONMENTAL IMPACTS OF LANDFILL DEVELOPMENT

**Environmental Impacts Associated with Landfill Development**

| Operational Phases of Landfill Development | Use of land | Highway congestion | Highway noise | Accidents | Noise | Air Pollution | Rats | Flies | Groundwater contamination | Economic loss | Reduction of Scenic Value |
|---|---|---|---|---|---|---|---|---|---|---|---|
| Haul to Landfill | | X | X | X | | | | | | | |
| Queue at Landfill | | X | X | X | | | | | | | |
| Excavation | | | | | | X | | | | | |
| Garbage Deposition | X | | | | | X | X | X | | X | X |
| Compaction | | | | | | X | | | | | |
| Soil Cover | | | | | | X | | | | | |
| Landscaping | | | | | | | | | | | |
| Leachate Generation | | | | | | | | | X | | |
| Leachate Treatment | | | | | | X | | | | | |
| Gas Generation | | | | | | X | | | | | |

21

# TABLE 3-4

## POTENTIAL ENVIRONMENTAL IMPACTS OF LANDFILL DEVELOPMENT

Use of land for excavation
Use of land for garbage deposition
Highway congestion haul to landfill
Highway congestion queue at land fill
Highway noise haul to landfill
Highway noise queue at landfill
Accidents due to haul
Noise from haul
Noise from compaction
Air pollution from garbage deposition
Air pollution from gas generation
Air pollution from leachate treatment
Rats from garbage deposition
Flies from garbage deposition
Groundwater contamination by leachate
Economic loss or gain from use of land
Reduction of scenic value

Construction of the matrix, and completion of the specific matrix boxes provides us with a complete list of potential environmental impacts as shown in Table 3-4.

With these impacts in mind we can now construct a list of sensitive resources. Again, some sub categories may be helpful in order to facilitate the listing process. Three useful categories are:

Natural Features
Landfill Resources
Land Use
Economic Factors

## TABLE 3-5

## RESOURCES SENSITIVE TO LANDFILL DEVELOPMENT

### Natural Features

Wetlands
Flood Plains
Surface Waters
Groundwater Resources
Fault Zones
Seismic Impact Zones
Unstable Areas
Expansive Soils
Subsidence Zones
Threatened & Endangered Species Habitat
Scenic Areas

### Landfill Resources

Sufficient depth of suitable soils for separation from groundwater resources
Sufficient depth of suitable soils for cover material
Existing depressions
Natural screening
Limited run on potential
Significant depth to groundwater resources
Minimal residential well density
Absence of groundwater conditions which make monitoring difficult
Minimal slope

### Land Use

Development (Existing or committed)
Airports (Jet or turboprop)
Prime farmland
Municipal water supply wells
Areas of historic importance
Areas of architectural importance
Areas of paleontological importance
Areas of archaeological importance
Areas with natural buffering
Proximity to areas with final use compatibility
Proximity to municipal boundaries
Areas of natural scenic beauty

TABLE 3-5

RESOURCES SENSITIVE TO LANDFILL DEVELOPMENT

CONTINUED

Economic Factors

Proximity to Major Highways
Highway restrictions
Traffic
Distance from centroid of waste generation or transfer
Availability
Land holding in large parcels

Table 3-4 and 3-5 are used to construct the second and final matrix as shown in Table 3-6.

## TABLE 3-6:
## MATRIX OF ENVIRONMENTAL IMPACTS OF LANDFILL DEVELOPMENT AND POTENTIAL SITING CRITERIA

**Environmental Impacts of Landfill Development**

| Natural Features | Use of land for excavation | Use of land for garbage deposition | Highway congestion, haul to landfill | Highway congestion, queue at landfill | Highway noise, haul to landfill | Highway noise, queue at landfill | Accidents due to landfill | Noise from haul | Noise from compaction | Air pollution from haul | Air pollution from garbage deposition | Air pollution from gas generation | Rats from garbage deposition | Flies from garbage deposition | Groundwater contam. by leachate | Economic loss or gain from use of land | Reduction of scenic value |
|---|---|---|---|---|---|---|---|---|---|---|---|---|---|---|---|---|---|
| Wetlands | X | X | | | | | | | | | | | | | | | |
| Flood Plains | | X | | | | | | | | | | | | | | | |
| Surface Waters | | X | | | | | | | | | | | | | | | |
| Groundwater Resources | | | | | | | | | | | | | | | X | | |
| Fault Zones | | | | | | | | | | | | | | | X | | |
| Seismic Impact Zones | | | | | | | | | | | | | | | X | | |
| Unstable Areas | | | | | | | | | | | | | | | X | | |
| Expansive Soils | | | | | | | | | | | | | | | X | | |
| Subsidence Zones | | | | | | | | | | | | | | | X | | |
| Threatened & endangered species habitat | X | X | | | | | | | | | | | | | | | |
| Scenic Areas | X | X | | | | | | | | | | | | | | | X |

TABLE 3-6, Continued

**Environmental Impacts of Landfill Development**

| Landfill Resouces | Use of land for excavation | Use of land for garbage deposition | Highway congestion, haul to landfill | Highway congestion, queue at landfill | Highway noise, haul to landfill | Highway noise, queue at landfill | Accidents due to haul | Noise from haul | Noise from compaction | Air pollution from garbage deposition | Air pollution from gas generation | Air pollution from leachate treatment | Rats from garbage deposition | Flies from garbage deposition | Groundwater contam. by leachate | Economic loss or gain from use of land | Reduction of scenic value |
|---|---|---|---|---|---|---|---|---|---|---|---|---|---|---|---|---|---|
| Depth of suitable soils for separation | | | | | | | | | | | | | | | | X | |
| Depth of suitable soil for cover | | | | | | | | | | | | | | | X | X | |
| Existing depressions | | | | | | | | | | | | | | | | | X |
| Natural screening | | | | | | | | | | | | | | | | | X |
| Limited run-on potential | | | | | | | | | | | | | | | | | X |
| Significant depth to groundwater resources | | | | | | | | | | | | | | | | X | |
| Minimal residential well density | | | | | | | | | | | | | | | | X | |
| Ease of groundwater monitoring | | | | | | | | | | | | | | | | X | |
| Minimal slope | | | | | | | | | | | | | | | | | X |

TABLE 3-6, Continued

**Environmental Impacts of Landfill Development**

| Land Use | Use of land for excavation | Use of land for garbage deposition | Highway congestion, haul to landfill | Highway congestion, queue at landfill | Highway noise, haul to landfill | Highway noise, queue at landfill | Accidents due to landfill | Noise from haul | Air pollution from haul | Air pollution from compaction | Air pollution from garbage deposition | Air pollution from gas generation | Rats from garbage deposition | Flies from garbage deposition | Groundwater contam. by leachate treatment | Groundwater contam. by leachate | Economic loss or gain from use of land | Reduction of scenic value |
|---|---|---|---|---|---|---|---|---|---|---|---|---|---|---|---|---|---|---|
| Development (Existing or Committed) | | | | | | | | | | X | | | | | | | | |
| Airports (Jet or Turboprop) | | X* | | | | | | | | | | | | | | | | |
| Prime Farmland | | | | | | | | | | | | | | | | | X | |
| Municipal water supply wells | | | | | | | | | | | | | | | | X | | |
| Areas of historic importance | | | | | | | | | | | | | | | | | X | |
| Areas of architectural importance | | | | | | | | | | | | | | | | | X | |
| Areas of paleonto- -logical importance | | | | | | | | | | | | | | | | | X | |
| Areas of archaeo- -logical importance | | | | | | | | | | | | | | | | | X | |
| Areas with natural buffering | | | | | | | | | | | | | | | | | | X |
| Proximity to areas with final use compatibility | | | | | | | | | | | | | | | | | X | |
| Proximity to municipal boundaries | | | | | | | | | | | | | | | | | X | |
| Areas of natural scenic beauty | | | | | | | | | | | | | | | | | | X |

* More correctly, there is a significant safety concern. See Chapter 4 for details.

TABLE 3-6, Continued

**Environmental Impacts of Landfill Development**

| Economic Factors | Use of land for excavation | Use of land for garbage deposition | Highway congestion, haul to landfill | Highway congestion, queue at landfill | Highway noise, haul to landfill | Highway noise, queue at landfill | Accidents due to haul | Noise from haul | Air pollution from haul | Air pollution from compaction | Air pollution from garbage deposition | Air pollution from gas generation | Rats from garbage deposition | Flies from garbage deposition | Noise from leachate treatment | Groundwater contam. by leachate | Economic loss or gain from use of land | Reduction of scenic value |
|---|---|---|---|---|---|---|---|---|---|---|---|---|---|---|---|---|---|---|
| Proximity to major highways | | | | | | | | | | | | | | | | | X | |
| Highway restrictions | | | X | | | | | | | | | | | | | | | |
| Traffic | | | X | | | | | | | | | | | | | | | |
| Distance from centroid of waste generation or transfer | | | | | | | | | | | | | | | | | X | |
| Availability | | | | | | | | | | | | | | | | | X | |
| Land holding in large parcels | | | | | | | | | | | | | | | | | X | |

28

From Table 3-6 we can then derive the final comprehensive list of siting criteria shown in Table 3-7.

## TABLE 3-7
## SITING CRITERIA FOR LANDFILL DEVELOPMENT

### Natural Features

Wetlands
Flood Plains
Surface Waters
Groundwater Resources
Fault Zones
Seismic Impact Zones
Unstable Areas
Expansive Soils
Subsidence Zones
Threatened & Endangered Species Habitat
Scenic Areas

### Landfill Resources

Sufficient depth of suitable soils for separation from groundwater resources
Sufficient depth of suitable soils for cover material
Existing depressions
Natural screening
Limited run on potential
Significant depth to groundwater resources
Minimal residential well density
Absence of groundwater conditions which make monitoring difficult
Minimal slope

### Land Use

Development (Existing or committed)
Airports (Jet or turboprop)
Prime farmland
Municipal water supply wells
Areas of historic importance
Areas of architectural importance
Areas of paleontological importance
Areas of archaeological importance
Areas with natural buffering
Proximity to areas with final use compatibility
Proximity to municipal boundaries
Areas of natural scenic beauty

## TABLE 3-7

### SITING CRITERIA FOR LANDFILL DEVELOPMENT
### CONTINUED

Economic Factors

    Proximity to Major Highways
    Highway restrictions
    Traffic
    Distance from centroid of waste generation or transfer
    Availability
    Land holding in large parcels

The reader will note that Table 3-7 is identical with Table 3-5. This is because we started with the known criteria in 3-5. In a typical application, Table 3-5 starts with many more criteria than are identified here, and the matrix in 3-6 is used to eliminate those criteria which are found to be of little or no significance.

## APPLYING THE CRITERIA

### The Phased Approach

Opposition to landfill siting is so intense and pervasive that any hint of a specific piece of property being seriously considered as suitable for landfill development seems to bring out heavy artillery. The serious planner must, however, complete all the necessary preliminaries in order to determine areas of greatest suitability. How can this investigative work be accomplished without putting the fear of landfill development into the region? The answer is the phased approach.

By careful phasing of the siting criteria already identified, it is possible to complete a significant portion of the siting work without specific site identification. In fact, a straightforward application of all the siting criteria selected would be impossible because many of the criteria are not capable of rendering in a yes or no format.

The most common basis of phasing is to select those criteria which are generally considered exclusionary as phase 1 and those which should properly be considered on a weighted basis as phase 2.

The term exclusionary in this context needs further explanation. For example, wetlands as an exclusionary item is readily understood. If wetland areas are mapped, then the area represented by the wetland designation should be considered as excluded. However, for areas with sufficient depth of suitable soils, the reverse is true. Here the exclusion zone is the area without sufficient depth of suitable soils.

The phased approach suggested comprises two phases:

        Phase 1 - Regional
        Phase 2 - Local

Regional Criteria

The regional phase suggested covers the following data:

TABLE 3-8
REGIONAL CRITERIA FOR LANDFILL SITING

Natural Features

        Wetlands
        Flood plains
        Surface waters
        Groundwater
        Suitable soils for groundwater protection
        Fault zones
        Seismic impact zones
        Unstable areas
        Expansive soils
        Subsidence zones

Land Use

        Development - Existing & Committed
        Airports
        Municipal Wells
        Prime Farmland

Economic Factors

        Proximity to Major Highways

The definition of each criterion is presented in Chapter 4, together with an explanation of why each feature is an exclusionary criterion.

It will be clear from perusal of this list that most of these features typically cover large areas. Mapping these items will therefore have the effect of excluding large areas from further consideration, thus making the second phase more manageable.

Local Criteria

The local siting phase covers the following data:

TABLE 3-9
LOCAL CRITERIA FOR LANDFILL SITING

Natural Features
        Depth of Suitable Soils for Cover
        Existing Depressions
        Natural Screening
        Run-on Potential
        Residential Well Density
        Ease of Monitoring Groundwater
        Slope
        Threatened and Endangered Species
        Scenic Areas
        Significant Depth to Groundwater Resources

Land Use
        Buffer Zone
        Final Use Compatibility
        Municipal Boundaries
        Area of Historic Importance
        Areas of Architectural Importance
        Areas of Paleontological Importance
        Areas of Archaeological Importance
        Highway Restrictions
        Traffic Impact
        Distance from Centroid of Waste Generation or Transfer
        Availability
        Land Holding in Large Parcels

The definition of each criterion is presented in Chapter 5 together with an explanation of why each feature is a design related criterion.

It will be clear from this list that most of these features are much smaller in scale than the earlier exclusionary criteria. Mapping each of these features over a wide area would be time consuming and costly. Once the regional criteria are mapped and specific areas can be targeted, many of these local features can be individually researched within these targeted areas with relative ease.

Further, it should be noted that many of these features are in the nature of a wish list as far as the designer is concerned. For example, the existence of a natural depression, natural screening or minimum run-on potential would be advantageous to the designer in terms of cost reduction. However, their absence is not insurmountable. Thus, the local criteria are more susceptible to analysis using weighting factors rather than the exclusion/inclusion method appropriate to regional criteria. Discussion of weighting strategies is to be found in Chapter 7.

# 4

# Regional Siting Criteria

## INTRODUCTION

We found in Chapter 3 that landfill site selection criteria could be derived by review of the various potential environmental impacts of the operation compared to the various sensitive environments likely to be encountered.

A selection of these selection criteria based on ease of mapping was proposed, the first set being described as regional site selection criteria. The regional criteria are as follows:

TABLE 4-1
REGIONAL CRITERIA FOR LANDFILL SITING

Natural Features

Wetlands
          Flood plains
          Surface waters
          Groundwater
          Suitable soils for groundwater protection
          Fault zones
          Seismic impact zones
          Unstable areas
          Expansive soils
          Subsidence zones

    Land Use

          Development - Existing & Committed
          Airports
          Municipal Wells
          Prime Farmland

   Economic Factors

          Proximity to Major Highways

Each of these criteria are mapped as exclusion zones. The composite of these maps is referred to as a sieve map because it sieves out the remaining areas once all the exclusion zones have been eliminated.

At this stage, each criterion will be discussed in greater detail in order to gain a fuller understanding of the role played in the siting process.

## NATURAL FEATURES

### Wetlands and Waters of the United States

Wetlands are an important natural resource. Wetlands are essential breeding, rearing and feeding grounds for a wide variety of fish and wildlife, including many threatened and endangered species. Many wetlands play an important role in flood protection and pollution control. The term "wetlands" means those areas that are inundated or saturated by surface or ground water at a frequency and duration sufficient to support, and that under normal circumstances do support, a prevalence of vegetation typically adapted for life in saturated soil conditions. Wetlands generally include swamps, marshes, bogs and similar areas.

The Federal Clean Water Act (CWA) (Public Law 92-500) addresses adverse effects upon wetlands by regulating impacts to "waters of the United States". The term "waters of the United States" means:

1.  All waters, including their adjacent wetlands, that are part of a surface tributary system to and including navigable waters of the United States (man-made, non-tidal drainage and irrigation ditches excavated on dry land are not considered waters of the United States under this definition);

2.  Interstate waters and their tributaries, including adjacent wetlands; and

3.  All other waters of the United States not identified in paragraphs (1) and (2) above, such as isolated wetlands and lakes, intermittent streams, prairie potholes, and other waters that are not part of a tributary system to interstate waters or to navigable waters of the United States if, in the opinion of the Division Engineer, the degradation or destruction of such waters could affect interstate commerce. The landward limit of jurisdiction in tidal waters, in the absence of adjacent wetlands, shall be the high tide line and the landward limit of jurisdiction in all other waters, in the absence of adjacent wetlands, shall be the ordinary high water mark.

    The term "navigable waters of the United States" means those waters of the United States that are subject to the ebb and flow of the tide shoreward to the mean high water mark and/or are presently used, or have been used in the past, or may be susceptible to use to transport interstate or foreign commerce.

The Subtitle D standards of the Resource Conservation and Recovery Act (RCRA) which are concerned with the disposal of municipal solid wastes identify the standards necessary to assure non-degradation of sensitive wetlands and waters of the U. S.

The protection of these areas is justified, by the fact that they make a number of important contributions to natural systems.

These areas store rainwater runoff, and reduce the velocity of flow from areas of high discharge. This has the effect of reducing the peak flood flow and, also, the frequency of flooding in downstream areas.

Depending upon the variety of species in the areas, the quality of the water that flows over and through them can be improved.

They provide food and habitat for many plant and animal species. They may significantly reduce shoreline erosion caused by wave action.

They can be recharge areas to groundwater systems supplementing local or regional groundwater flow by infiltration or percolation of surface water.

These critical environments can be adversely effected in several different ways by the location of landfills in or near these sensitive areas. The primary concerns of waste disposal of waters of the U. S. or wetlands are the physical disturbance and the potential for discharge of waste or discharge of leached material into sensitive eco-systems.

The deposition of solid waste in these areas destroys sensitive vegetation or results in advancement of a succession of dry land species by blocking normal flow. This may have the effect of reducing or eliminating nutrient exchange by reducing the productivity of the existing system and altering the species which are dependent upon the existing pattern for survival.

The U.S. Army Corps of Engineers administers the section 404 permit programs under the Clean Water Act (CWA Section 404) which controls filling wetlands or waters of the U. S. In practice the U. S. Army Corps requires the presence of three specific indicators before it will accept that an area meets the definition of wetland:

TABLE 4-2

CRITERIA FOR WETLAND IDENTIFICATION

| | |
|---|---|
| 1. | Hydrophytic Vegetation |
| 2. | Hydric Soils |
| 3. | Wetland Hydrology |

The absence of any one of these will typically lead to a designation as "waters of the United States."

As a practical matter the U. S. Army Corps administration of protection is essentially the same whether the area is a wetland or waters of the United States. The protection of wetland, however, is generally regarded as a higher priority.

As far as developing data for mapping siting criteria is concerned, both areas need to be treated alike since the problems created by siting landfills in wetlands or waters of the U. S. are equally significant.

In most states, the Fish and Wildlife Service of the U. S. Department of the Interior maintains the most comprehensive wetland maps available. It should be noted that these maps are based on earth satellite photo interpretation as well as data on file and as such some technical interpretation is necessary. However, the mapped data does identify all known wetlands and waters of the U. S. which is what is needed in this phase of mapping regional criteria. At a later stage in the process if no site is found free of wetland or waters of the U. S., then it may be necessary to go back to this data and make a determination of just how significant each area is.

Flood Plains

The potential for floods to increase the risk of contamination from landfills is obvious. The risks range from increased leachate production to a complete washout of landfill facilities. The destructive force associated with floods depends upon the depth, velocity, and duration of the flood. The amount of debris in floodwaters can significantly add to the erosion potential of floodwaters. Since landfills are simply containers for a wide variety of solid waste, an inundated landfill significantly increases downstream erosion. The time for implementation of emergency measures aimed at protecting the facility or removing wastes to a safe location, is very limited. Once contaminants enter surface waters, transport is generally rapid, and the contaminants migrate rapidly to downstream waterways.

By their nature, floods typically cover large areas, as a result of which large areas are subject to contamination where landfills are impacted. In coastal or estuarine areas, tidal effects disperse contaminants upstream and downstream of the source. Mixing and dispersion dilute contaminant concentrations. Contaminants can become adsorbed to stream sediments and remain long after soluble and suspended contaminants have been swept away. Suspended contaminants tend to settle in slow moving portions of a river or stream or they may spread along the coast. This spread of contaminants can produce long term contamination problems that can adversely affect downstream water supplies and aquatic resources many years after the source of contamination has been removed.

Contaminants in sediments can destroy benthic organisms, or enter the benthic food chain, resulting in toxic effects higher in the food chain. Contamination of surface waters and sediments, due to releases from landfills, can destroy aquatic organisms directly through toxic effects, including food chain accumulation of low level concentrations. Contaminated sediments can destroy spawning grounds for fish and shell fish and other benthic invertebrates which inhabit or feed in the sediments. Fisheries can be damaged, due to elimination of their benthic food supply, or through consumption of contaminated organisms. Destruction of spawning habitat may limit the reproductive success of fisheries. Other indirect effects include changes in species composition or diversity when sensitive species are eliminated and replaced by less desirable pollution tolerant species. In severe cases, all organisms in the vicinity of the release may be destroyed.

Floodplain areas are defined conventionally on a statistical basis using the return frequency of the flood. For example, a 100 year floodplain is defined as that area likely to be subject to floods with a return frequency of once every hundred years. A 100 year floodplain is the area most commonly designated by a variety of federal agencies, since the data is readily available to the designer in the form of 100 year flood maps.

One hundred year floodplain maps are generally available from the following sources:

State Flood Control Agencies

Federal Emergency Management Agency (FEMA)

Local and Regional Planning and Zoning Agencies

Soil Conservation Service - U.S. Department of Agriculture

U.S. Army Corps of Engineers

National Oceanic and Atmospheric Administration

Federal Housing Administration (FHA)

U.S. Geological Survey

Bureau of Land Management - U.S. Department of the Interior

Bureau of Reclamation - U.S. Department of the Interior

River Basin Commissions and Special Flood Control Districts

Local and State agencies involved with public works construction

At the level of regional mapping, all 100 year flood zones should be mapped as exclusion zones. Only in the event that no area can be found outside the flood zone should it be necessary to return to this data and assess the potential for flood-proofing as an alternative approach.

Surface Waters

The release of solid wastes to surface waters may also result in acute and chronic human health problems. The primary route to exposure is through drinking water. However, the consumption of fish and other aquatic organisms that accumulate chemicals is also a concern, since human health effects may occur at lower levels of bioaccumulation relative to those that effect indigenous aquatic organisms. Recreational use of surface water for water contact activities can also result in contaminant exposure.

A major concern associated with the contamination of surface waters is that existing ambient water quality criteria for human toxic and carcinogenic protection are in many cases below detection limits in water. Potentially toxic releases to surface waters from landfill facilities can therefore go undetected. In addition, the human health effects of chronic exposure may not be exhibited for decades.

Surface waters represent an economic asset to regional and local areas. Contaminant releases to surface waters, that result in widespread long term degradation of water quality, can also effect agricultural and industrial use.

In order to map this particular criterion a suitable separation distance is generally specified. Two hundred feet is often used because this distance provides a reasonable space in which to construct a protective dike or containment structure to isolate the surface water body from the potential effects of the landfill. In regions where there are significant differences in elevation between the projected site and the surface waters, a greater separation distance may be necessary.

Maps of major water bodies are typically available from a wide variety of sources, only a small selection of which are listed here:

US Geological Survey
NOAA
Local and Regional Planning and Zoning Agencies
Commercial Map Companies
Water Resource Agencies
River Basin Commissions

## Groundwater

The source of groundwater is precipitation. Both rain and snow infiltrate loose particles of the soil and eventually percolate into the ground. Below a certain depth, called the zone of saturation, or water table, almost all openings in geologic material are filled with water. Above the water table, these openings, or pore spaces, are filled with both water and air. This definition of water table is not related to the availability of groundwater to wells. A tightly packed, fine grain material may be completely saturated with water, yet the yield and rate of recharge would not be sufficient for use.

Groundwater is stored in the zone of saturation. In this zone, groundwater is stored in openings ranging from tiny spaces between particles of clay and silt, to large gaps in sand and gravel, and crevices in dolomite and limestone. The pore space in any earth material is its porosity expressed as a percentage of the total volume of the material. The size and interconnection of pores determine how easily water moves through material from areas of high potential energy to areas of low potential energy.

The property of the material that describes the ease of water movement through it is called the hydraulic conductivity or permeability. The rate of groundwater and contaminant movement depends upon the hydraulic conductivity, the hydraulic gradient, and the effective porosity of the material. An aquifer is a body of earth materials yielding enough water to a well to satisfy the need for drilling it. So an aquifer supplying adequate water for a single resident might not be an aquifer for a municipality.

Aquifers may be unconfined or confined. In an unconfined aquifer, the water table is the top of the aquifer. No impermeable materials overlay the aquifer confining it. In a confined aquifer, also known as an artesian aquifer, the groundwater is confined under pressure greater than atmospheric pressure by overlying relatively impermeable materials. This pressure causes the water in a well to rise above the top of the aquifer, this is called an artesian well. If the water in a well rises to the surface, it is a flowing artesian well.

The existence of groundwater on its own is not necessarily a cause for exclusion from landfill siting. More information is needed about the groundwater in order to identify its relative importance.

We will see later in this chapter that more information is needed about the type of landfill itself, because in certain circumstances, landfills may be specifically designed to handle high groundwater conditions. Such landfills are called Inward Gradient Landfills. However, in the first phase of landfill siting, inward gradient conditions are not preferred criteria. If, in the final analysis, no conventional site can be found, then and only then, is it necessary to return to the siting criteria and determine if an inward gradient site can be found.

The additional information about the groundwater required for a siting study falls into several categories:

Existing Groundwater Quality
Shallow Groundwater
Existing Use

Before we embark on a discussion of the variety of forms of groundwater data potentially available, it is important to review again the reason for mapping regional criteria in the first place. Briefly, we need to reduce the area of detailed (and costly) search to manageable proportions. Thus, if any of the above forms of data are readily available, they are recommended for use as regional criteria. However, if data in any one of these areas is available for only limited areas of geography, then they are best left until later.

Existing Groundwater Quality

Groundwaters generally have a wider range of variability in natural chemical quality that is normally found in surface waters. Some groundwaters are of better quality than minimum standards normally accepted for drinking water, and may be used without treatment. Ground waters of exceptional quality represent an extremely valuable resource. Other groundwaters contain high quantities of dissolved substances, and these resources may be so costly to treat they are virtually unusable. Other groundwaters have radioactive constituents which emanate from geologic materials that generally do not impact surface waters.

Groundwater quality may be classified in several different ways. In urbanized areas, groundwater quality may be identified as a function of existing use. Where existing aquifers have been rejected on the grounds of groundwater quality, however, it cannot be assumed that these areas are not at risk from further contamination. Battles rage on both sides of this question. On the one hand, the pressures of increased development result in the need for increased water supply and this may place demands upon existing groundwater sources that can only be met by exploiting groundwater sources which were thought previously to be too highly contaminated to warrant treatment. On the other hand, good quality groundwater can be pumped a considerable distance before the cost of pumping is offset by the cost of treating a nearby contaminated water supply. There is no doubt that areas of high quality groundwater should be one of the criteria for protection. Whether areas of contaminated groundwater should also be mapped for future use depends upon circumstances.

If large areas of potential use are identified in the mapping process without specifically utilizing contaminated groundwater data, it is advisable to leave areas of poor groundwater quality until later in the search. As we will see in the next chapter, proximity to areas of contaminated water may later be utilized as a factor in weighing the merits of one option compared to another.

Map information depicting groundwater quality is still sparse. In most states, information on groundwater quality is only available for major aquifers which have been the focus of study for specific purposes. Nevertheless, this information is invaluable when it is available and every effort should be made through specialized state agencies to identify any known groundwater quality data in the areas under study.

Shallow Groundwater

All groundwaters are potentially at risk from landfill leachate. Groundwaters are protected from the risk of contamination by liners (natural and/or manmade). Liners add a degree of protection, because they retard leachate movement long enough to remove it for treatment. Landfill designers seek as much back-up protection as possible by searching for areas where there is sufficient depth of soils with low permeability above the groundwater to provide a substantial amount of natural protection. As we will see later, this condition of natural protection is generally considered under the category of suitable soils. For our purposes, under this category, provided that only areas protected by suitable natural soils are mapped, then areas of shallow groundwater resources will be protected. This condition provides for ample opportunity to monitor the geologic material under the landfill for contaminants and take corrective action before the groundwater is impacted.

Depth to groundwater data is readily available in most states.

For mapping purposes, shallow groundwaters (groundwaters less than 20-30 feet from the surface) should be mapped as exclusion zones.

## Existing Use

All groundwater subject to existing use should be protected. Many states have enacted groundwater protection legislation to formalize the need for protection.

In the absence of any specifically legislated protective zone, it is suggested that all groundwater recharge zones be protected where the groundwater is currently utilized.

A different protection zone is recommended for municipal well heads, and this will be discussed later under municipal wells.

## Suitable Soils for Groundwater Protection

For sanitary landfills, perhaps the single most important criterion is the availability of suitable soils. Soil suitability for the landfill designer falls into two categories:

1. Cover Material
2. Liner Material

Landfills are defined as areas of solid waste where the waste is covered daily with six inches of compacted daily cover. Some exceptions are permitted in rural areas which do not operate every day. This soil cover must be workable by heavy equipment and cohesive enough to form a uniform six inches of compacted cover over solid waste at the end of each days filling. Most soils are capable of fulfilling this purpose, except those at the extremes of soil classification where the preponderance of sand, silt or clay is so high that the soils are difficult to work with.

Landfills also fill with water which is called leachate, and this leachate will tend to move through the soil unless it is retarded by confining layers. Modern landfills seek to reduce the risk of leachate loss by dependence upon one or more liners to retard the flow. Liners may be of 2 kinds; synthetic or natural. Natural liners should be composed of dense clay with field permeabilities of $1 \times 10^{-7}$ cm/second or less.

Suitable soils for liners are the silty or clay tills. For sanitary landfills, material of this kind is ideal if it has no areas interbedded with sand and gravel and is readily accessible upon excavation prior to filling.

Since suitable liner soils are the most critical for siting landfills, they are commonly the criterion of choice. In addition, suitable liner soils can usually be rendered viable as cover soils by mechanical working, and/or by intermixing additives such as sand or lime to make them more workable.

Thus, the most appropriate criterion for mapping purposes is soil having the same characteristics as a good liner or better (i.e. permeability of $1 \times 10^{-7}$cm/sec or less).

In geological terms the soil which separates any underlying ground water from direct infiltration from above is called the confining layer. In some states information is already mapped which identifies the depth of the confining layer. The choice of which depth to identify as the appropriate siting criterion is dependent upon availability of data. In several states where the search for landfill space is actively supplemented by state geological institutions, data is already available on soils with permeability less than $1 \times 10^{-7}$cm/sec at a depth of 50 feet or lower. This is ideal for mapping purposes with the caveat that, if in the course of mapping this information little or nothing remains as suitable, the next step is to back up, and either reduce the minimum thickness allowed, or raise the required permeability.

Fault Zones

Earthquakes are usually caused by movements of tectonic plates along faults. Faults are fractures in rock along which the adjacent rock surfaces are displaced. Faults may vary in length from a few feet to several hundred feet. The presence of faults indicates that, at sometime in the past, movement along the fault has occurred. Movement may be either a slow slip, which produces no ground motion, or a sudden rupture resulting in perceptible ground motion that is known as an earthquake.

Movement along the fault plain is called a slip. Faults assume different geometric forms. Faults may be classified into three types; normal, reverse and strike slip. A normal fault has movement down the dip of the rock, while a reverse or thrust fault has movement up the dip of the rock. Vertical displacements seen at the surface are produced by both normal and reverse faults. A strike slip fault has a relative displacement, as viewed from the ground surface, that is at right angles to the dip. This type of fault is sometimes referred to as a wrench or tear fault.

When the displacement along the fault occurs, movements of great masses of material develop dynamic effects. The dynamic effects initiate vibrations within the earth's crust. These vibrations, or seismic waves, travel for great distances in all directions. It is because of these waves that a localized movement of the earth can have a disastrous effect upon a wide area. The investigation of sites for the possible hazard of surface fault rupture is a difficult geologic task. Many active faults are complex, consisting of multiple breaks. The evidence for identifying active fault traces is generally subtle or obscure, and the distinction between active and long inactive faults may be difficult to make.

The California Division of Mines and Geology has developed a guideline for evaluating the hazard of surface fault rupture. The guideline outlines suggested measures for the detection, evaluation, and investigation of surface and near-surface faults. Fault rupture associated with earthquakes can cause significant damage to a landfill. Damage may include the rupture of the leachate collection and liner system. Failure of the leachate collection system can prevent the removal of leachate, thereby allowing the leachate to pond on the liner. If the liner system is ruptured, this can create a pathway for leachate to migrate into and contaminate the uppermost aquifer. In addition to the potential damage to the leachate collection and liner system, the integrity of the landfill slopes could also be impaired by fault rupture, resulting in the potential for exposing solid waste to surface runoff.

A fault zone can generally be divided into a main fault zone, a branch fault zone and a secondary fault zone. The relationship of individual faults in the fault zone has resulted in this classification system. The main fault zone contains the main fault, that is, the fault with the greatest displacement. Occasionally, faults diverge from and extend well beyond the main zone of faults, and are referred to as branch faults.

Secondary faults are completely separate, spatially, from the main fault, and sometimes several hundred feet to a few miles from the main fault. Associated with main, branch or secondary faults are often small subsurface faults evident as fault plains running parallel to the fault, and typically are considered a part of that fault. Adjacent to the fault rupture is commonly a zone of deformation. This is an area where the ground has been bent or warped as a consequence of the two surface plains moving relative to one another.

Surface deformation is frequently reported within a zone of several tens to several hundred feet wide. Structures located within this zone are subject to distortion and likely to be subject to damage.

Landfills located in an area of fault rupture and surface deformation are liable to significant damage. In general, it is recommended that new landfills should not be located within 200 feet of a fault that has had displacement in Holocene time (the Holocene is a geologic time unit that extends from the end of the Pleistocene to the present, and includes approximately the last 11,000 years). The best protection for landfills is to avoid faults subject to displacement.

Fault zones may be mapped based on geologic maps, which are generally available from:

    U.S. Geological Survey
    State Geologic Agencies

Seismic Impact Zones

The movement of great masses of the earths crust naturally develop dynamic effects that emanate vibrations, or seismic waves, within the crust. The seismic waves travel for great distances in all directions, such that a localized movement of the earth can have a disastrous effect over a wide area (Legget and Karrow, 1983).

Earthquakes may cause the partial or total collapse of buildings, bridges, man-made and natural slopes, and other systems. Failure of the structures may be the result of their inability to withstand the ground motion or of the foundation subgrade collapsing under seismic loading.

Regional seismicity maps have been available since the 1950's to aid the engineer in designing buildings and other structures to reduce the effects of earthquakes. Regional seismicity or risk maps usually do not attempt to reflect geologic conditions, nor take into account, variations in soil properties. The maps are useful primarily to provide insight into the relative hazard across the United States, together with results concerning the relative importance of the various parameters involved. (Algermission and Perkins, 1976; Bolt et al, 1975). A single parameter, such as a probabilistic estimate of maximum acceleration, does not provide all of the information necessary to describe all the characteristics of strong ground motion that are important in the design of structures. Nevertheless, a wide range of structures can and have been designed to be earthquake-resistant, using peak acceleration as the basic ground motion data (Algermission and Perkins, 1976).

The United States Geologic Survey has published an open file report addressing the probabilistic estimate of maximum acceleration in rocks in the contiguous United States. The report presents maps of the relative earthquake hazard in various seismic zones of the continental United States based on a constant probability level. Maps provided with the report identify the maximum horizontal accelerations in hard rock expressed as a percentage of the earth's gravitational pull, with a 90% probability that it will not be exceeded in 10, 50 and 250 years (Algermission and Perkins, 1982).

Another major cause of destruction during an earthquake is the failure of the ground structure by loss of strength that occurs in loose saturated sandy soils. This phenomenon, termed liquification, is a result of an increase in pore water pressure. The increase in pore water pressure decreases and sometimes eliminates the shear strength of the soil bodies of loose relatively fine uniform sand below the water table which are susceptible to liquification during an earthquake, especially if the duration of the quake is long enough for the occurrence of a large number of oscillations involving repeated reversals of shearing strains of large amplitude. Soil that has lost all shear strength behaves like a viscous fluid. Liquification often appears in the form of sand boils during earthquakes. When a soil fails in this manner, a structure resting on it simply sinks (Terzaghi and Peck, 1967).

Liquification causes 3 types of ground failure:

> Lateral Spreads
> Flow Failures
> Loss of Bearing Strength

Lateral spreads involve the lateral movement of large blocks of soil as a result of liquification in the subsurface layer.

Flow failures consisting of liquified soil or blocks of intact material riding on a layer of liquified soil, are the most catastrophic type of ground failure caused by liquification.
When the soil supporting a building or structure liquifies and loses its bearing strength, large deformations can occur within the soil allowing the structure to settle (USGS, 1981).

Ground motion associated with earthquakes damages landfills. Damage may include failure of dikes and berms, resulting in exposure of solid waste. Also, ground motion can severely rupture leachate collection and liner systems. Runoff can be contaminated by contact with exposed solid waste, thereby creating the potential to contaminate surface water. Failure of the leachate collection system may prevent the removal of leachate, thereby allowing the leachate to pond on the liner. If the liner is ruptured, this may cause a pathway for leachate to migrate into and contaminate the uppermost aquifer.

Failure of natural and man-made slopes adjacent to the landfill may also affect the facility. Failure of these slopes may damage run-on and run-off control systems, leachate disposal, and slope management systems, as well as reducing the stability of the containment structures.

The State of California requires that containment structures be capable of withstanding the maximum probable earthquake. The maximum probable earthquake, as defined by the California Division of Mines and Geology, is the maximum earthquake that is likely to occur during the 100 year interval.

California guidelines require that the following be considered when deriving the maximum probable earthquake (California, 1975):

The regional seismicity considering the known past seismic activity

The fault or faults within a 100 km radius that may be active within the next 100 years

The types of fault

The seismic reoccurrence factor for the area and faults (when known) within the 100 km radius

The mathematical probability of seismic activity associated with the faults within the 100 km radius

The postulated magnitude should not be lower than the maximum that has occurred within historic time

Earthquakes can affect landfills through ground motion, surface faulting, earthquake induced ground failures, and even tidal waves. Although earthquakes cause much less economic loss annually in the United States than ground failures and floods, major earthquakes have the potential to cause sudden and considerable loss.

The U.S. EPA has proposed that all new landfills located within a seismic zone with a 10% probability that the maximum horizontal acceleration and hard rock will exceed 0.10g in 250 years must have all containment structures including liners, leachate collection systems and surface water control systems designed to resist the maximum horizontal acceleration for the seismic zone.

This requirement translates to a 4% probability of exceeding the maximum horizontal acceleration in 100 years.

<u>Unstable Areas</u>

Unstable areas include landslide prone regions, areas with comprehensive or expansive soils or ultrasensitive (quick) clays, subsidence prone areas and Karst terrains.

Landslide Prone Areas

Landslide is a general term covering a wide variety of mass movement land forms and processes involving the downslope transport of soil and rock material under gravity. Landslides are a significant hazard in virtually every state in the United States. Although individual landslides are not as spectacular or as costly as other geologic and hydrologic hazards, they are more widespread.

Landslides can be classified on the basis of the type of movement and the type of material involved. The types of movement are:

Folds
Couples
Slides
Spreads
Flows
Combinations of two or more of the above

The two classes of materials that are involved in landslides are:

Bedrock
Soils consisting of debris, earth or a combination of the two

All landslides involve the failure of earth materials under shear stress. Any human activity or natural event that increases shear stress or lowers shear stress can trigger a landslide.

The major causes of landslides are:

Construction Operations or Erosion
Earthquakes and Vibrations
Rains or Melting Snow
Freezing and Thawing
Dry Spells
Seepage from Man-made Sources of Water

Construction operations or erosion can cause a landslide if they affect the support of the slope. Slides are common in excavated cuts for highways, railways, quarries, and pits. Erosion of the toe of a slope can leave the remaining slope face unsupported and subject to sliding. Heavy buildings located close to the edge of a slope can initiate a slide.

Earthquakes and vibrations from blasting or construction related operations can cause spontaneous liquefaction of loose sand, silt or loess deposits situated below the groundwater table. Under similar circumstances, sensitive clays can undergo a decrease in shear strength. If the structure of these deposits collapses, pore pressures increase and approach the total overburden pressure. At this point, the shear strength of the soil is drastically reduced, and the soil, behaving like a heavy liquid, flows downhill along with the overburden. Rains or melting snow can cause increased pore water pressures leading to reduced shear strength along the potential slip faces. Most slope failures occur after heavy rains, or during spring snow melts, when large quantities of water penetrate cracks and fissures.

Freezing or thawing can induce cracking in rock formations resulting in rock slides. In silty soils, the freeze/thaw cycle can cause increased pore pressures and ground surface movements. The drying of soils may result in crack formation when overburden pressures are reduced as a result of reduction in soil water content. These cracks can cause the shear strength to be reduced, increasing the risk of slides. Drying may also lead to the development of shrinkage cracks in shales, particularly in exposed formations.

Areas in the Appalachian Mountains, Rocky Mountains, and the coastal ranges along the Pacific Ocean have the most severe landslide problems. All types of slope movement occur in these areas. Large areas in the mid-continent region are underlain by relatively weak shales, and in these areas landsliding is prominent on moderate to steep slopes. When the underlying materials are weak, large excavations commonly produce landslides, even in flat areas. Large parts of Alaska and Hawaii are also severely affected by landslides.

Expansive Soils

Certain types of soils and soft rocks will expand when they become wet and shrink when they dry out. Swelling or expansive soils are generally rich in clay minerals, particularly, smectites (montmorillonites). These minerals swell by absorbing water that enters and expands the space between the plates which constitute the crystalline structure of the clay mineral. The process is reversible, and upon drying, the spacing decreases and the clay shrinks.

Montmorillonite clays are the most prone to this behavior, with bentonite clay being an extreme case that can increase in volume by a factor of 10 as it goes from dry to saturated state.

Expansive soils are widely distributed. Smectites are the most abundant in geological formations throughout the Rocky Mountains, the upper Great Plains, the southern Gulf Coast Plain, and along the Pacific Coast. They are also locally abundant throughout the Great Basin Region and along the Atlantic coast.

The amount of expansion that can occur depends on the type and quantity of clay mineral present, and it is a function of time, confining load, initial density and initial water content. Expansive clay formations that in nature do not exhibit excessive shrink/swell activity can become a problem as a result of construction activities that remove overburden or allow greater access of water to the formation.

Subsidence

Subsidence is defined as "the lowering or collapse of the land surface either locally or over broad regional areas" (USGS, 1981). Subsidence can be the result of either natural or human activities. Natural causes of subsidence include dissolution of limestone and other soluble materials, earthquakes, and volcanic action.

Limestone and dolomite are slightly soluble in water. In hot, wet climates, the solution process can cause voids to form in formations of these minerals. If overlying materials collapse or subside into the solution cavities, a surface depression called a sinkhole forms. Where limestone or dolomite deposits are widespread and sinkholes are common, the land surface is referred to as Karst topography.

Sinkholes vary in depth from slight inundations to over 100 feet deep. Typically, depths range from 10 to 30 feet, with areas in the range of a few square yards to several acres. There are two major classes of sinkhole. The first type forms from the collapse of an underground void. The more common type, the doline, develops slowly by the downward solution of the underlying rock. Surface water entering the first type of sinkhole tends to flow rapidly underground through outlets called swallow holes. Runoff draining into a doline usually percolates slowly underground through the soil to the bottom of the sinkhole. If a sinkhole becomes clogged, a temporary pond or lake forms. If the outlet is unplugged, rapid drainage can occur.

Karst Terrain

Karst terrain can be found throughout the United States. Sinkholes are to be found in parts of southeastern and midwestern states.

The removal of solid materials from underground deposits can also result in subsidence. This is particularly true of shallow coal mines, where adequate roof support was not provided and collapse occurs during mining or long after closure.

Solution mining can also cause subsidence. In solution mining, water soluble minerals, such as salt, gypsum, or potash, are dissolved and pumped into an underground formation where the material to be mined is taken into solution. The solution rises to the surface where the water can be evaporated and the material recovered. Huge underground cavities are formed by this activity causing surface subsidence often many years after mining. With some soils and soft rock formations, the addition of water can cause subsidence.

One type of formation where this happens is called loess. This material has the characteristics of soft rock since it is formed from fine grain materials and cemented together in an extremely loose open structure by a water soluble mineral cement. In the dry state, loess is firm and hard and serves as a foundation material. When subject to excessive wetting from sprinkling or ponding on the surface, the mineral cement is dissolved and the soil structure collapses.

In landslide prone areas, construction on natural slopes could cause problems if the placement of waste overburden leads to a situation in which the shear strength of the underlying materials is exceeded. It is also possible that vibrations generated by construction equipment on the site could trigger ground movement. In unstable areas where liners are constructed, it is important to avoid removing the toe of the adjacent slopes that could slide into the landfill and damage the liner. In areas with expansive clays, liners can be torn, and sidewalls can collapse if uneven expansion and bulging takes place below the liner. In areas that experience constant subsidence problems associated with the presence of underground cavities, the risk that subsidence would occur below a landfill is significant. This is particularly true if the weight of the solid waste exceeds the bearing strength of the materials supporting the landfill.

The resolution of many of these problems is expensive, and the results less than 100% reliable.

Avoidance of these areas is the best course of action, and these areas are easily plotted as exclusion zones.

## LAND USE

### Development - Existing & Committed

The public perception of landfills is such that some level of separation must be established between land use development and landfill. However, the distance criterion itself is difficult to define.

A distance of 2000 feet separation has been shown to be unreasonable[1], since over this distance such natural obstacles as rivers, hills, or mountains are capable of completely isolating the supposed risk. On the other hand, distances of the order to 500 feet appear to be more reasonable. Over this distance, noise from a working facility is reduced to acceptable levels and most other potential impacts can be effectively screened.

---

[1]"Economic Impact of Prohibiting Landfill Development Within 2000 Feet of Public Schools" by Noble & Associates, Inc. for Illinois Institute of National Resources February 1979.

However, even 500 feet should be handled with reserve in densely developed areas. In these areas, recognizing that design parameters need adjustment to meet special circumstances, even 500 feet may not be enough.

Land use development maps are available in a wide variety of designations. In some areas, development maps are available which only identify development as one category regardless of whether it is residential, commercial, or industrial and for reasons explained later, these may be used.

In some areas, maps depicting specific land use categories are available. Typical categories are:

Residential
Commercial
Industrial
Institutional
Recreation
Open Space

In these cases, exclusion zones may be more readily identified for each category based upon specific separation distances. Here we run into problems, because the residential separation distance which presumably represents what the public will accept when they are at home may not be the same as that proposed for the public when they occupy commercial or industrial establishments. Where there are differences, the criteria selection team should be particularly careful to document the reasons for these differences. With the possible exception of industrial development, it is generally advisable to adopt the same separation distance for all categories of land use.

For industrial development in certain areas, there may be good reason for assuming a lesser separation distance such as 250 feet.

In discussing development as a siting criterion, we need to address not only different types of development, but also, whether the development is existing or potential. Existing development is usually available as a mapped resource. Potential development, on the other hand, can mean different things to different people and may not be available in any mapped resource. There is a risk that potential development, if loosely defined, may even be manipulated as a means of excluding landfill development. There is no fixed rule on how potential development should be defined, since it will vary from one area to another.

In one recent study, potential development was defined as committed development, which had received final plat approval, and which had been active within the last two years. The rationale for this choice might help to establish the parameters selected in other areas.

Basically, plat approval was taken as signifying a specific level of commitment to a development. However, in some cases in this particular study, area activity on the land in question had lapsed for a considerable length of time and this was taken to indicate that although the original development was serious, it had lapsed and that the area could be considered for landfill. A lapse of two years was taken to be a reasonable cut-off period.

It is important to note that existing and potential development is not the same as zoning. In many states, zoning is specifically denied as a restriction on landfill development. The use of zoning to exclude landfill is ultimately self defeating, since it can be so easily manipulated to exclude all potentially suitable areas. In the long term, exclusion tactics by any one community or region will lead to retaliation by potential host communities.

## Airports

On October 16, 1974, the Federal Aviation Administration (FAA) issued Order 5200.5 in response to an increasing number of bird strikes at airports near landfills. In summary, the Order recommends that landfills located within 10,000 feet of a runway which is used by jet aircraft, or within 5,000 feet of a runway which is used by propeller driven aircraft, should be closed. The Order also recommends the closing of a landfill where the flight path of the aircraft crosses the route taken by birds in travelling to or from the landfill to or from water or roosting areas.

The U.S.EPA standards for regulating landfills adopted the same restrictions as identified in FAA Order 5200.5.

In 1988, these same restrictions were cited by U.S. EPA in the proposed siting criteria for sanitary landfills. Eleven states and territories specify the same location restrictions and two specify more stringent standards.

The distance restriction is measured from the end of the runway, as shown in the Figure 4-1 diagram. Where both runways handle the same type of aircraft, the restricted areas overlap to form an irregular circular shape. Where runways handle different types of aircraft, the larger circle will obviously govern as shown in Figure 4-2.

It should be noted that although gulls are the species most often implicated in bird/aircraft collisions in the United States, collisions with a wide variety of birds have been documented. (Blokpoel 1976; Forsythe 1974; Ultrasystems Inc., 1977).

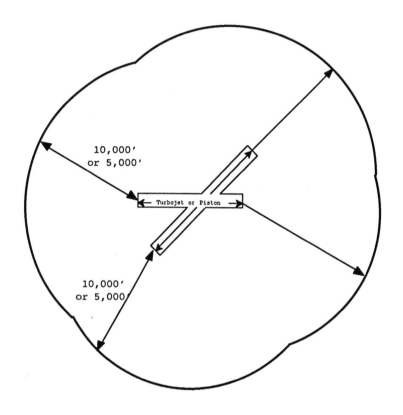

10,000'
or 5,000'

Turbojet or Piston

10,000'
or 5,000'

**Figure 4-1 : Measurement of Exclusion**
**Distance from Runway**

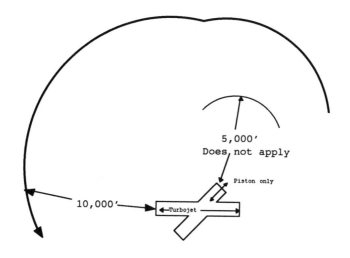

A. Runways Handling Different Types of Aircraft

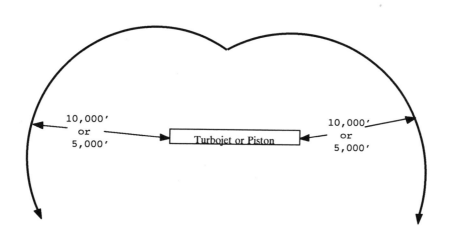

B. Single Runway, Either Type of Aircraft

**Figure 4-2:**
**Measurement of Exclusion Distances from Runway**

A large proportion of bird strikes occur below altitudes of 500 feet, and since aircraft are generally below this altitude within the distances specified by the FAA order, this restriction should prevent most bird/aircraft collisions.

However, in some areas of the country where the bird population in a specific region is particularly high, the siting criteria could be more stringent and a larger radius of exclusion could be used.

On January 31, 1990, FAA Order 5200.5 was cancelled, and 5200.5A was instituted. The new order is to be found in Appendix A.

In most respects the restrictions on landfill siting remain the same.

Specifically, under the criteria section, the order reads as follows:

"Disposal sites will be considered as incompatible if located within areas established for the airport through the application of the following criteria:

a.   Waste disposal sites located within 10,000 feet of any runway end used or planned to be used by turbine powered aircraft.
b.   Waste disposal sites located with 5,000 feet of any runway end used only by piston powered aircraft.
c.   Any waste disposal site located within a 5 mile radius of a runway end that attracts or sustains hazardous bird movements from feeding, water or roosting areas into, or across the runways and/or approach and departure patterns of aircraft."

However, a "zone of notification" was added to the criteria in order to provide the appropriate FAA Airports office with an opportunity to become involved in the selection process.

The relevant section reads as follows:

"Additionally, any operator proposing a new or expanded waste disposal site within 5 miles of a runway end should notify the airport and the appropriate FAA Airports office so as to provide an opportunity to review and comment on the site in accordance with guidance contained in this order. FAA field offices may wish to contact the appropriate State director of the United States Department of Agriculture to assist in this review. Also, any Air Traffic control tower manager or Flight Standards Office manager and their staffs that become aware of a proposal to develop or expand a disposal site should notify the appropriate FAA Airports office."

<u>Municipal Wells</u>

Although groundwater is addressed earlier as a specific site selection criterion, it is important to note that some specific features of groundwater use may need additional protection.

Areas of groundwater recharge to municipal wells are a prime example. Depending upon the volumes being pumped, and the cone of depression, significant areas of land may need to be avoided in the proximity of municipal wells.

A maximum of a 1000 foot radius is suggested, but in special circumstances, a greater radius may be used.

Municipal well locations may be obtained from state water resource organizations.

<u>Prime Farmland</u>

Prime farmland is essentially land which has the best combination of properties to promote growth. However, such land is also a function of a great many economic considerations which can ultimately be defined as demand.

The physical properties of prime farmland are defined in the following table:

TABLE 4-3
PHYSICAL PROPERTIES OF PRIME FARMLAND

1. The soil must be on gentle slopes and not subject to excessive erosion.

2. Soil permeability must be at least 0.15 cm per hour in the upper 50 cm.

3. Gravel, cobbles, and stones must not be so common as to interfere with modern farm tractors and equipment.

4. Soil depth to hardpan or bedrock must be great enough to minimize interference with adequate water storage or normal root extension.

5. Soil pH must be suitable for the crop to be grown and the salt and sodium content must be acceptable.

6. The soil should not be subject to excess water from flooding or a high water table.

7. Soil moisture should be adequate and dependable from normal rainfall or from irrigation.

8.    The mean annual soil temperature at a depth of 50 cm must be greater than 0 °C and the mean summer soil temperature greater than 8 °C for soils with an O (organic) horizon and 15 °C for other soils.

The selection of prime farmland as a site selection criterion should be handled with great care. While many soils in the area may conform to the above physical properties, the economic factors noted earlier may not be conducive to development as farmland. Thus areas which may have all the characteristics of prime farmland may be considered as suitable for landfill development in cases where there is no significant demand for the use of the land for farming.

## ECONOMIC FACTORS

### Proximity to Major Highways

In order to minimize the cost of landfill development, it is advisable to keep the amount of new access road development to a minimum.

The selection of the actual minimum distance should be handled with discretion, because too short a distance will significantly reduce the area of suitability, while too great a distance will result in too large an area to assess in more detail later.

Any new landfill will require major road development, regardless of proximity to the highway. It is important in determining the distance to the highway to include the site road itself, otherwise, land area needed for the site will be excluded.

Since the landfill design is not developed until the site is selected, the assessment of site road length must be an estimate. A useful start for this estimate is to assume a square plan, and a 20 foot depth of excavation. It is also necessary to estimate the life of the site. A later section of this chapter provides all the necessary formulae to estimate the overall area needed. The length of site road is equal to the length of one side of the square.

An additional 500 feet is suggested as the maximum length of extra road necessary to gain access to the site from the highway. For example, if the maximum desired site is to provide for 2000 TPD, the site road distance is 3200 feet. Add 500 feet for additional site road, and 3700 feet is the distance from the major highway which must be mapped.

The selection of a definition for "major highway" is also an important consideration, since whatever highway is selected must be capable of withstanding heavy vehicular wheel loads typical of packer trucks or transfer trailers. The appropriate wheel loads are shown in Figure 5-5.

Once the appropriate distance is selected and the correct highways are identified based upon wheel loading, the road maps can be obtained from a variety of sources.

The selected highways are extracted from the available information, and the suitable area is defined as a broad strip on either side of the highway (the highway itself should, of course, be excluded).

## SIEVING

The derivation of areas having high potential for landfill siting is comparatively straightforward.

Based upon the mapping of the exclusion zones identified earlier, it only remains to sieve out the remaining areas left over once the exclusion zones are eliminated.

As noted earlier, this may be done by the preparation of maps on acetate sheets, or by using computer mapping techniques. While the acetate mapping technique is perfectly acceptable, it should be clear at the outset that the procedure is significantly less flexible than computer mapping. On the other hand, the initial set-up cost in terms of computer hardware, not to mention the sophisticated software necessary to overlay computer maps, can be prohibitive to smaller communities.

However, computer mapping is in use in almost every state in the U. S., and a great many cities and counties already possess extremely sophisticated geographical information systems. In the mapping of the regional criteria, it may be possible in some states to employ state or county expertise for a fraction of the cost of outright purchase of the necessary equipment and expertise.

Alternatively, many planning and environmental consultants also possess sophisticated computer mapping equipment and expertise, and this resourse also represents a viable alternative to purchasing the equipment and building up expertise from scratch.

Whatever method is used, it is important that decision makers fully understand the derivation of the selection criteria. It is for that reason that the matrices used in the determination of the criteria are incorporated into this book. Use these tools wisely, and they will repay the time spent many times over. The use of a rational system for site selection does not eliminate criticism, and the more the user understands the basis of site selection, the better the criteria can be defended when the time comes to present the plan.

## THE SIEVING PROCESS

The sieving process is very straightforward. The mapped criteria are basically arranged one upon the other, and the areas not included by any of the criteria are suitable areas for landfill. The production of a single sieve map from several criteria maps is shown in Figure 4-3.

# FIGURE 4-3

## CRITERIA MAPPING TO PRODUCE A SINGLE
## SIEVE MAP

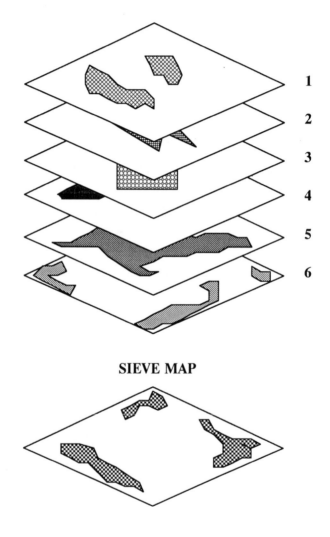

**SIEVE MAP**

As noted earlier, the exclusion zone may be the criterion itself, such as wetlands where the mapped feature is a positive image of the criterion. Alternatively, the mapped feature may be the reverse of the criterion itself where the mapped feature is a negative image. An example of a negative feature is proximity to major highways where the aim is to minimize the length of additional road necessary to gain access to the landfill. The area mapped is based upon a measured distance from all major highways, but the exclusion zone is everything beyond the line drawn.

The differences in the mapping of these two criteria are illustrated in Figure 4-4.

**FIGURE 4-4**

**DIAGRAM SHOWING POSITIVE AND NEGATIVE
EXCLUSION ZONES**

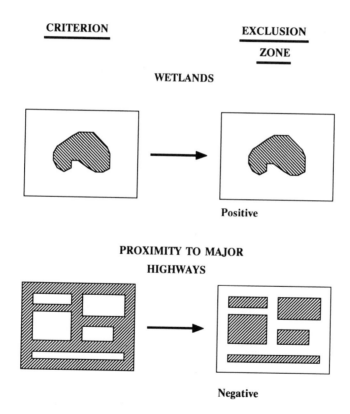

**CRITERION**

**EXCLUSION**

**ZONE**

**WETLANDS**

Positive

**PROXIMITY TO MAJOR
HIGHWAYS**

Negative

## THE REGIONAL SITING MAP

The final product of the sieving process is a regional siting map. In most instances this map identifies areas of real estate which taken together represent a significantly reduced area of study.

The remaining area represents the first sieve of areas with potential for landfill siting. It should be stressed that the delineation of this remaining area is simply the first step in the siting process. While you have succeeded in reducing the area which will now be examined in greater detail, you should not have succeeded to the degree that those living in the area can immediately identify a specific site. The reason should be obvious. If only one site can be identified at this stage, then the more detailed process of applying the local site selection criteria will almost certainly take place in such a blaze of publicity that it may be difficult at times to apply the criteria with the appropriate degree of scientific detachment.

We will proceed on the assumption that the regional process has produced a map rather similar to that shown in Figure 4-5. At this stage it is necessary to discuss the difficult problem of sizing.

# FIGURE 4-5

## A TYPICAL REGIONAL SITING MAP

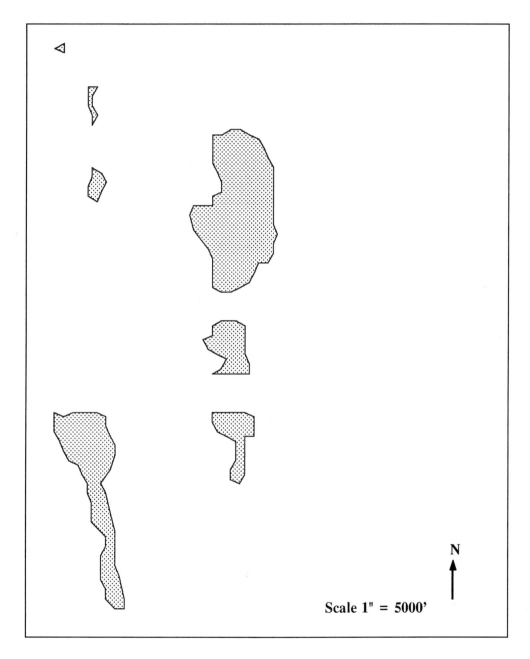

Scale 1" = 5000'

N

# SIZING

Stated simply, the first step is to decide how long the landfill is to last. Given the time and expense involved in siting, the general view is that this should be a long time, but it is important that the search should not exclude all reasonable options.

It is necessary to identify the minimum size in order to reduce the time spent looking at small parcels only to find that development is uneconomical.

The computation of minimum size begins with an assessment of total population served. This should be comparatively straightforward.

The weight of solid waste to be landfilled is determined by the use of a waste generation factor derived from local or national sources. This generation factor is an approximation of the amount of solid waste generated per capita per day.

The following formula can be used to determine the total annual waste generated per ton per year.

$$\text{Total Annual Waste} = \frac{365 \, P \times W \, (1 - f)}{2000} \quad \text{Tons per year}$$

Where:  P  = Population
  W  = Waste Generation Factor (Pounds/Capita/Day)
  f  = Recycle Factor (%/100)

Landfilled waste takes up a certain amount of air space depending upon the compaction, and the compaction depends upon the technology used to compact the waste. The following table will assist in identifying the necessary compaction density to use:

### TABLE 4-4
### COMPACTION DENSITIES

| COMPACTION METHODS | COMPACTION (Pounds/Cubic Yard) |
|---|---|
| Tracked Dozer | 750 - 900 |
| Toothed Wheel Compactor | 850 - 1100 |
| Low Density Baler | 1000 - 1250 |
| High Density Baler | 1400 - 1750 |

The following formula will establish the annual in-place volume of the waste.

In Place Volume $= V_{IP} = \dfrac{2000A}{Cd}$ Cubic yards/year

Where: $A$ = Total Annual Waste (tons)

$Cd$ = Compaction Density (pounds/cubic yard)

It is important to allow for the volume of daily cover in this equation. Air space must be found for the in-place volume of the solid waste, and also the compacted cover required over every cell constructed each day.

Air Space Volume $= V_1 = V_{IP} (1 + C/12h)$ cubic yards/year

Where: $V_1$ = Air space requirement for one year
$V_{IP}$ = In place volume (cubic yards/year)
$C$ = Cover depth (inches) (Six inches in most states)
$h$ = Daily cell height (feet)

This formula will provide a good approximation of the air space needed for one year of operation. However, except under exceptional circumstances, it is a poor use of resources to spend a year or more planning and implementing a siting strategy for a site which will only last for one year.

A twenty year planning period for a landfill is common throughout the United States. The twenty year period has been justified in the past on the grounds of savings in scale. The point being that the fixed costs of building access roads, ticket office, weigh scale, and administration buildings have to be financed during the life of the site, and that these costs are high compared to the landfill itself. It is important to realize that this logic is rooted in the historic notion that the landfill itself was basically a hole in the ground for the long term storage of waste. Compared to the fixed costs of the ancilliary requirements, the fixed cost of the hole in the ground was comparatively modest. This imbalance has changed significantly in the last 5 - 10 years. Increasingly landfill is becoming a complex treatment process in which the most harmful contaminant (leachate) is extracted for further treatment in a procedure which in some cases does not end until certain leachate standards are met.

The fixed costs of preparing the impoundment area for leachate collection in addition to the preparation of the final cap itself which is now a sophisticated combination of gas channel, infiltration seal, and plant bed are now considerable. So much so that they dwarf the ancilliary fixed costs referred to earlier.

Simply stated, most of the investment in modern landfill goes into the hole in the ground. Thus the savings in scale argument is no longer valid.

A second argument against the twenty year planning period is that the rate of growth in urban America is rapidly consuming those large tracts of land necessary to provide disposal volume for such a long period. A decision to go for a twenty year site is often a decision in favor of high transportation charges, high gasoline consumption, and higher air pollution.

Finally, I have found that one of the most potent arguments against landfills throughout the country is that no one wants to live with them for twenty years. This argument is often coupled with the complaint that what happens after twenty years is simply guesswork, and the prospect of a further expansion cannot be discounted. Far better in my view to plan for five years with a guarantee that the landfill will be landscaped and closed for good when that time is up.

Let us assume, therefore, that the minimum air space we seek is as follows:

$$V_{1-5} = 5 \, V_{IP} \, (1 + C/12h) \text{ cubic yards}$$

Where: $V_{1-5}$ = Air space requirement for five years

If necessary population projections and recycle rates can be altered incrementally in this equation by returning to the derivation of the total annual waste. In this case, the in-place volume requirement would vary from year to year, and $V1-_5$ would be the sum of all five computations.

$$V_{1-5} = V_1 + V_2 + V_3 + V_4 + V_5$$

Remember, however, that we are still in the regional siting stage, and a first level approximation is perfectly adequate. The air space requirements, for example, are more complex than represented in the above equation because we have yet to account for the volume taken up by the leachate collection system and the final cap.

In order to produce a convenient model of the area needed, assume we are looking for a square configuration. It will be clear as we proceed that a square is simply a useful conceptual tool with which to estimate suitability of certain areas of study.

At this stage we need to find the area of land needed to provide $V_{1-5}$ cubic yards (the volume of air space needed to take 5 years worth of waste).

The equation for the volume of the frustum of a pyramid is needed to solve this problem as shown in Figure 4-6.

## FIGURE 4-6

## VOLUME OF THE FRUSTUM OF A PYRAMID

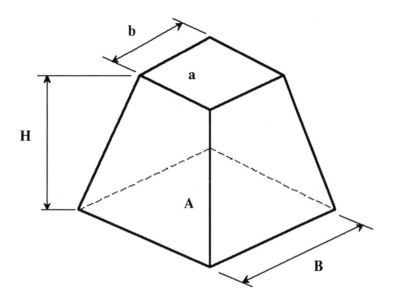

$$\text{VOLUME} \quad V \ = \ \frac{H\,(A + a + \sqrt{Aa}\,)}{3}$$

Where :      A = Area of Base
a = Area of Top
H = Height

Or :    VOLUME    $V \ = \ \dfrac{H\,(B^2 + b^2 + Bb\,)}{3}$

Where the base and top are square
and

B = Length of side at base
b = Length of side at top

A typical landfill approximates two frustums, one below, and one above grade as shown in Figure 4-7.

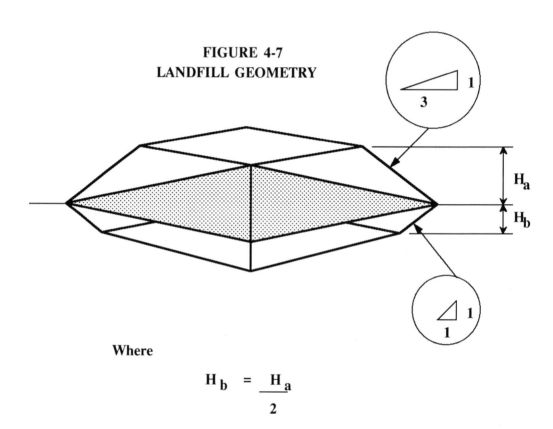

**FIGURE 4-7**
**LANDFILL GEOMETRY**

**Where**

$$H_b = \frac{H_a}{2}$$

Assuming the daily and final cover is to be supplied by the excavated material, and only granular material will be imported to the site, it is a useful approximation to use the one third/two thirds rule.

This rule states that the landfill depth below ground level will be one third the total depth of the landfill, and the height will be two thirds. This being the case, we can simplify the total volume.

$$V_a = \frac{H_a (B^2 + b^2 + Bb)}{3}$$

Since the slope above grade is 1:3

$$b_a = B - 6H_a$$

$$V_a = H_a/3 \; [B^2 + (B - 6H_a)^2 + B(B-6H_a)]$$

$$V_a = H_a (B^2 + 12H_a^2 - 6BH_a)$$

$$V_a = B^2 H_a + 12 H_a^3 - 6BH_a^2$$

Similarly,

$$V_b = \frac{H_b (B^2 + b^2 + Bb)}{3}$$

Since the slope below grade is 1:1:

$$b_b = B - 2H_b$$

$$V_b = H_b/3 \; [B^2 + (B - 2H_b)^2 + B(B - 2H_b)]$$

$$V_b = H_b/3 \; (3B^2 + 4H_b^2 - 6BH_b)$$

But $\quad H_b = H_a/2$ According to the one third/two thirds rule

$$V_b = H_a/6 \; (3B^2 + H_a^2 - 3BH_a)$$

$$V_b = \frac{B^2 H_a}{2} + \frac{H_a^3}{6} - \frac{BH_a^2}{2}$$

$$V = V_a + V_b = 1.5B^2 H_a + 12.167H_a^3 - 6.5BH_a^2$$

$$V = 1.5 Ha (B^2 + 8.11H_a^2 - 4.34BH_a)$$

$$V = 1.5 H_a [B^2 + 8.11H_a (H_a - 0.535B)]$$

---

Thus knowing the height above grade ($H_a$), and the dimension of the base (B), we can arrive at the total volume (V). Remember that this total volume (V) is to be made equal to the generated volume of waste.

Very often the overall height above grade is a prime concern depending upon the topography in the area. Thus in situations where $H_a$ and the five year volume are known, the dimensions of a square lot can be easily calculated.

Assuming the application of the above formula reveals an area requirement of say 50 acres, we can then return to the regional sieve map and exclude areas too small for further consideration. An area of 50 acres is approximately equal to an area of 1500 x 1500 feet, or at the scale shown in Figure 4-5, a square measuring 3/10 of an inch on each side. This area is shown in Figure 4-8.

With this modified regional sieve map, we are now ready to proceed with the local site selection.

# FIGURE 4-8

## A TYPICAL REGIONAL SITING MAP
## SHOWING HOW SMALL AREAS ARE EXCLUDED

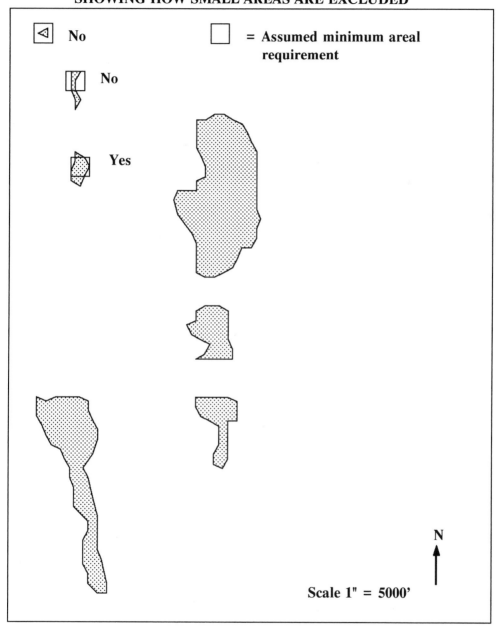

# 5

# Local Siting Criteria

## INTRODUCTION

As noted earlier, there are a large number of criteria left over after applying regional siting criteria, most of which are not exclusionary. These criteria are best handled as design considerations, and as such they must be weighted accordingly. The various methods of weighting will be discussed in a later chapter.

The local siting criteria are presented in the following table:

TABLE 5-1
LOCAL CRITERIA FOR LANDFILL SITING

Natural Features

    Depth of Suitable Soils for Cover
    Existing Depressions
    Natural Screening
    Run-on Potential
    Residential Well Density
    Ease of Monitoring Groundwater
    Slope *
    Threatened and Endangered Species *
    Scenic Areas
    Significant Depth to Groundwater Resources

Land Use

    Buffer Zone
    Final Use Compatibility
    Municipal Boundaries
    Areas of Historic Importance *
    Areas of Architectural Importance *
    Areas of Paleontological Importance *
    Areas of Archaeological Importance *
    Highway Restrictions
    Traffic Impact
    Distance from Centroid of Waste Generation
    Availability
    Land Holding in Large Parcels

*These items are also exlusionary, but owing to difficulties in data collection, they are best applied as local criteria.

At this stage we will review each criterion and the features which contribute to a full understanding of how it can be weighted for further consideration.

## NATURAL FEATURES

### Depth of Suitable Soils For Cover

Cover is absolutely vital to landfill operation. Cover is necessary to control, vector breeding and animal attraction, water movement into the fill, gas movement out of the fill, fire hazard, litter, odor, unsightliness, assist vehicular support and provide a base for future landfill development.

In order to meet these many demands, a cover material must conform to certain minimum requirements.

Soil is primarily made up of sand, silt and clay. The percentage distribution of each of these constituents in addition to the particle sizes, determines the texture. Variations in the distribution of the three primary constituents and the general terminology for soil type can be shown very clearly in a soil classification chart or ternary diagram as indicated in Figure 5-1.

# FIGURE 5-1

## SOIL CLASSIFICATION CHART

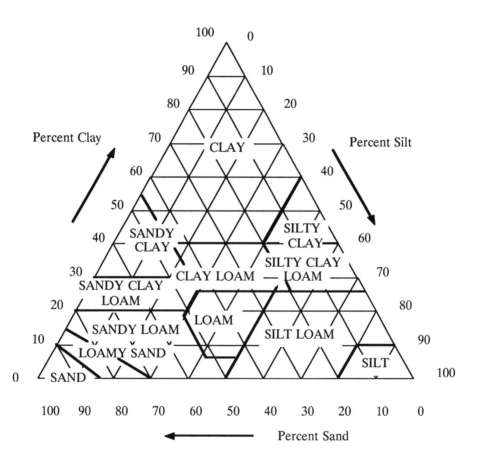

In general, soil with good workability and compaction characteristics is the most desirable cover material. A soil containing large quantities of clay may present an operational problem during wet weather and, under certain conditions, may tend to shrink and crack, thereby making it difficult to maintain a tight cover. Clayey soils may also present a difficult excavation problem during cold weather.

Sands and loamy sands are too permeable and non-compactable, and, to a lesser degree, similar problems occur with pure silts. Therefore, the mixed loams with low clay contents have been determined to be the best cover material.

Selection of a loam as the most desirable cover material has been recommended in a number of standards for sanitary landfill, ASCE Manual No. 39 recommends a soil containing 50 to 60 percent sand, with the balance being approximately equal amounts of clay and silt. A U.S. Public Health Service publication on landfills states that sandy loam is considered to be an excellent cover material as is a sandy loam or sandy silt (up to 50 percent silt). Pennsylvania Department of Environmental Resources generally considers the mixed loams with low clay contents as the most suitable cover material, in general agreement with the ASCE (1959), U.S. Public Health Service (1968) and the New York State Department of Health (1969).

Some controversy surrounds the boundary lines between acceptable and unacceptable cover materials and there is considerable divergence of view when it comes to including the clay loams and the silty and sandy clays. Where there is strict regulation at a state level, this should be followed. However, the notion that certain soil types should be excluded on the grounds that they will give trouble in working takes no account of the mechanical means available to modify soil workability. Where clay loam, for example, is readily available, a good case can be made for using it if a good friable soil material is scarified into the loam. To prevent cracking upon drying when used as final cover, the addition of 3 to 6 inches of top soil will keep the surface moist and prevent drying.

In addition to areas rich in mixed loam, areas rich in clay can also be used as cover soils by layering the final cover. As shown in Figure 5-2 the clay can be compacted as an effective barrier after which it is covered by 6 - 18 inches of loam to provide a good basis for vegetation and also to keep the clay moist, and free from cracks which appear in clay when the moisture content is reduced.

To prevent the clay from cracking from gas pressure buildup under the cap gravel can be used as a gas channel under the clay.

## FIGURE 5-2

## LAYERED LANDFILL CAP

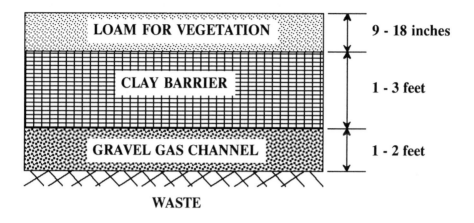

On balance a wide range of soils may be used as cover material provided the appropriate precautions are taken. The ternary diagram can be modified to show the soils suitable for cover material as shown in Figure 5-3.

# FIGURE 5-3

## SOILS SUITABLE FOR COVER MATERIAL

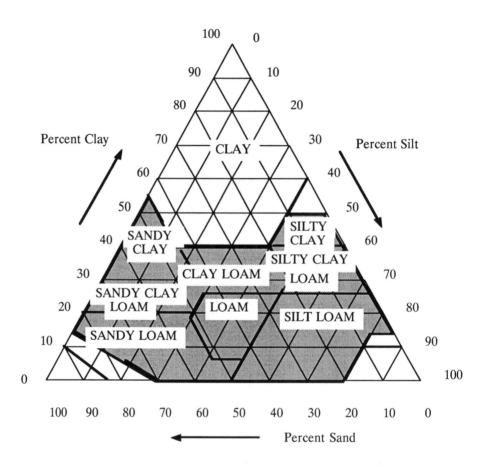

## Existing Depressions

Areas of existing depression are a potentially valuable resource in the search for a landfill site. In the language of the landfill designer, they already represent usable air space.

Mapping this particular resource requires some knowledge of the area needed for the landfill, because the size of the existing depression must bear some reasonable relationship to the size of the projected landfill.

For example, if 200 acres are required, there is little value in plotting areas of existing depression of 10 acres or less. The value of such a small area in the overall design is limited. On the other hand, existing depressions of 100 acres or more would be potentially valuable resources.

Existing depressions may take the form of manmade or natural depressions. Manmade depressions are more likely to yield good prospects for landfill siting simply because natural depressions are most frequently formed by water flow or unstable soil conditions, both of which are features to avoid in landfill siting.

Man-made depressions, on the other hand, are often designed to exclude water flow from adjacent drainage, and in this regard, such sites may be good candidates.

Man-made depressions caused by quarrying for sand and gravel need to be considered on their merits. Extensive remaining deposits of sand and gravel in the area will reduce consideration of such sites as their increased permeability will eliminate them in the regional selection. On the other hand, if all sand and gravel have been removed, and a sound clay base remains with only minor lenses of sand and gravel, such sites may be feasible provided they have not become a safe haven for threatened and endangered species in the interim. It should be remembered that limited availability of cover material may also be a significant factor in this type of site.

Sandstone or limestone quarries are less likely candidates for landfill unless it can be shown that the remaining bedrock is sparsely fissured, and the underlying bedrock is not water bearing. Again, the need for cover as well as liner material will greatly impact the overall cost of this type of application

## Natural Screening

Screening is necessary to reduce noise, but also, general public overview. While public access for conducted tours of landfill should be encouraged, public viewing from homes or highways should be as limited as possible, since once seen, landfills tend to be perceived as generating noise and odor problems, even when they are not the guilty party. Screening also tends to reduce wind speed, and thus, minimize wind blown litter and transmission of odors.

Both in the development of the landfill design itself and, also, to some extent in the final use plan, it will be necessary to develop a screening plan. The quantity of natural screening already in existence in a particular area will, therefore, raise its value as a potential site.

This is not simply an economic factor in the sense that trees, shrubs and topographic variation are already there so they do not have to be paid for, but also a significant psychological factor. A site area currently screened is perceived as an area less disturbed by landfill activity than one which is in full view, where the screening has to be planned and implemented as the fill progresses.

Screening as a selection factor is difficult to map, and care should be taken in the mapping process to map not only the area which is occupied by screening material, but also the area screened.

Run-On Potential

An important factor in the management of a modern sanitary landfill is the management of rainfall run-off. Considerable planning, earthmoving and diversion structure construction is necessary to successfully prevent water from the surrounding area (or run-on) from encroaching upon the area of the fill. Naturally, this whole exercise is easier and less costly if the quantity of run-on to be managed is at a minimum.

A watershed map will easily identify the direction and size of watersheds in the region. A watershed map is shown in Figure 5-4 as a series of interconnected areas. Each area represents a drainage basin or a sub-drainage basin in a large river valley, which may cover thousands of acres in some areas. Arrows are used to represent the predominant flow direction. The important consideration for a potential site is its relative position in the watershed, and the relative size of the area upstream of the site.

This particular map is used as an overlay. As shown in Figure 5-4, site A would be more desirable than site B in this particular example, because site B is located further down stream than site A, which means that the site, if constructed, would have to manage much greater volumes of run-on compared to site A.

# FIGURE 5-4

## TYPICAL WATERSHED MAP

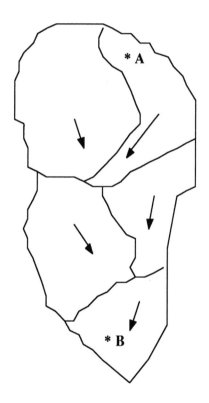

## Residential Well Density

Modern landfill design includes groundwater monitoring devices designed to monitor groundwater quality and ensure protection of potable water supply.

In the event that contamination shows up in monitoring wells, extensive remediation is necessary to ensure that contamination does not spread beyond the boundary of the site.

If the remediation is not fully successful, local groundwater wells are most at risk from further contamination. In the final chapter in this book, a variety of strategies are discussed for improving prospects of siting, one of which is water quality protection. The days of simply giving broad public assurance that water quality will not be affected by a new facility are over. New facilities will be increasingly required to enter into contractual agreements with local residents within a certain radius of the site to guarantee water quality and provide assurance against impairment.

Under these circumstances, it is better to grant groundwater quality assurance to a small number of residential well owners than a large number. Thus, residential well density is another comparative criterion. Zero would be best and low numbers more acceptable than high numbers. As the reader will learn later, it is proposed that the landfill operator should not only guarantee groundwater quality, but also test on a quarterly basis. Given the high cost of analysis, limitations in the number of wells to be tested also has some significant advantages.

## Ease of Monitoring Groundwater

Groundwater monitoring is crucial to the day-to-day success of any landfill. It is critical that monitoring data should be easily understood, and not so complex that it is subject to a wide variety of different interpretations. Ease of interpretation requires an absence of complex groundwater regimes, and an absence of other features in the area which could be sources of contamination in any way similar to leachate.

Thus, any areas of complex hydrogeology should be mapped as exclusion zones. Similarly, existing sources of contamination, i.e., old landfills, tanks, or old industrial sites with known contamination and the groundwater systems they may impact, should also be excluded.

## Slope

Extensive areas with a natural slope in excess of 1 vertical to 3 horizontal are a potential problem for the landfill designer for several reasons;

1. High run-on potential
2. Risk of soil erosion, and/or slope failure
3. High energy costs for equipment operation
4. Limited air space in relation to plan area
5. High cost of run-on management

The matter of high run-on potential has already been covered by the earlier mapping. Here it is necessary to map extensive areas with steep slopes as exclusion zones.

## Threatened and Endangered Species

Sightings of threatened and endangered species are generally available from state conservation departments and from the Fish and Wildlife Division of the Department of the Interior. These data need to be supplemented by maps of nature preserves, parks and other areas of preservation, since the sightings themselves do not generally define habitat.

The limited space devoted to this criterion should not detract from its importance. Threatened and endangered species habitat should be excluded as far as landfill siting is concerned. The only reason it is not typically mapped as a regional criterion is that regional mapping of this criterion is rarely available.

Often sightings themselves do not define habitat with great precision, and the work involved to establish habitat can be considerable. It is more appropriate therefore to exclude threatened and endangered species habitat when specific areas of interest are defined.

It is important to note that threatened and endangered species are defined both at a federal and a state level. In addition many species are "under consideration" prior to listing and it is prudent to identify species in this category as well as those which are listed.

## Scenic Areas

Known scenic area should be mapped in order to determine whether potential sites are subject to overview from these areas.

It is important to note that scenic area overview is a concern only to the extent that the view of the proposed landfill cannot be mitigated. This might be true, for example, where a scenic area at a higher elevation than the landfill looks down upon the operation. In this situation, any attempt to screen the landfill would probably be ineffective, particularly if the difference in elevation is substantial.

Keep in mind that screening need not be limited to areas around the landfill. A single view or residence high on a hill can be screened by tree planting near the view itself rather than at the landfill site.

## Significant Depth to Groundwater Resources

Since groundwater is generally considered to be the bane of landfill design, it may seem strange that depth to groundwater appears here in the list of local siting criteria, and not in the first round of criteria. However, to some extent a groundwater criterion is implied in the regional criteria by the choice of a factor such as depth of suitable soils for groundwater protection.

Having isolated areas with low permeability soils, however, it is now necessary to establish the depth to groundwater in these soils. Groundwater may exist in these soils under a hydrostatic head, under these conditions the groundwater level is referred to as a piezometric level.

In general, a low piezometric level is more desireable than a high level, simply because the natural soil barrier between the base of the landfill and the low level is greater in the former case than the latter. In this situation, in the unlikely event that a leak occurred, the thicker natural barrier in the former case allows more time for remedial action.

However, areas where the piezometric level is high (or close to the projected base of the landfill) should not necessarily be discarded. In recent years, a design technique known as inward gradient design has evolved which allows for the possibility of using sites hitherto considered unsuitable. Inward gradient design has definite potential for some areas, but it is more costly than conventional design, and in the final analysis, more leachate is generated by inward gradient systems and this too increases operating costs. Accordingly, inward gradient prospects should be considered low priority compared to more conventional designs.

## LAND USE

### Buffer Zone

Most states require a buffer zone around landfills, but even where there is no regulatory requirement, it makes good sense to provide one. A buffer is not necessarily the same as screening. A buffer zone must be contiguous with the landfill, and represents a strip of land which will always separate the landfill from any adjacent land uses.

Where a buffer zone is also a natural screen, the screening area already mapped should be remapped wherever it is immediately adjacent to a potential site. This information can then be used to establish a comparative weighting for one site compared to another.

### Final Use Compatibility

Modern landfill practice generally dictates open space as a final use, or at least some development compatible with extensive open space. The responsibilities and liabilities associated with post closure care of landfills are such that the integrity of the landfill cap precludes almost any form of development except open space.

There are few land uses which would be incompatible with open space. However, at the other end of the scale there are some land uses where the development of associated open space, recreational or otherwise, would be a positive benefit. Thus, the degree or extent of compatibility will need to be weighted in order to provide a method of comparison.

## Municipal Boundaries

Municipal boundaries are an important criterion primarily as a location finder. Major highways may also be used for this purpose, but in some remote areas they may not be as helpful as municipal boundaries.

In some circumstances, the siting of a landfill specifically inside or outside a particular minicipality can be a matter of concern for jurisdictional reasons. A municipality may want to exercise municipal regulation over a facility, and this would not be possible if it lay in an unincorporated area.

## Area of Historic Importance

The National Historic Preservation Act of 1966 established the National Register of Historic Places, created the Advisory Council on Historic Preservation and the State Historic Preservation Programs, and provides Federal funds for the preservation of historic properties.

Section 106 of the Act and implementing regulations (36 CFR Part 800) require Federal agencies (the Regional Administrator in the case of new landfills) to evaluate the impact of new activities on properties listed or eligible for listing in the National Register of Historic Places, and to adopt measures, when feasible, to mitigate any potential adverse effects where determined necessary by this evaluation. In addition, this evaluation must be implemented in cooperation with the State Historic Preservation Officers and, when appropriate, in consultation with the Advisory Council on Historic Preservation.

Most states have their own version of the Historic Preservation Act, and in most cases, there is now a State Register of Historic Places. In many states, county or local communities also have their own version of preservation orders and, where applicable, these designations should be observed in the development of site selection criteria.

At a federal level, there is also the Wild and Scenic Rivers Act. The purpose of this Act was to establish a national wild and scenic rivers system that includes selected rivers of the nation which, with their immediate environments, possess outstanding scenic, recreational, geologic, fish and wildlife, historic, cultural or other similar values. Under this Act, selected rivers are to be preserved in their free-flowing condition, and are to be protected along with their immediate environments for the benefit and enjoyment of present and future generations.

Section 7 of the Act prohibits Federal agencies from assisting, by license or otherwise, the construction of any water resources project that would have a direct, adverse effect on a national wild and scenic river or its immediate environment.

The distance limitation for buildings of historic or architectural significance varies from state to state. Many states give no guidance on assessing impact, and leave the assessment to the State Historic Preservation Officer. Owners of historic property who may be adversely disposed toward landfill siting tend to take a long view when determining how far their viewshed extends. A distance of 1000 feet is suggested.

## Areas of Architectural Importance

In many communties specific areas of architectural importance have been established by state, county or local regulation. These areas are often recipients of special grants to aid in rehabilitation or other aspects of protection. As in the case of historic areas, these areas need special protection in order to preserve their unique character. An exclusion distance of 1000 feet is suggested.

## Areas of Paleontological Importance

Paleontology is the study of former geological epochs by studying fossil remains. Areas where fossils have been found are mapped by State Historic Preservation Offices, or in some cases, Departments of Natural Resources.

This information is often jealously guarded due to the risk of valuable sites being overrun by unskilled treasure seekers. In order to map this kind of information, it may be necessary to explain in detail that point specific information is not vital, and that broad general areas of exclusion are acceptable. An exclusion distance of 250 feet is suggested.

## Areas of Archaelogical Importance

Archaeology is the study of ancient times by studying physical remains of art, and cultural implements. As in the case of paleontological finds, this information is often difficult to obtain from the relevant agencies.

Commonly, an explanation of the purpose of the exercise (i.e., to exclude development in such areas) is enough to obtain some measure of cooperation. The fact that individual maps showing these areas are rarely part of the public document should also assist in obtaining cooperation.

An exclusion distance of 250 feet is suggested.

## Highway Restrictions

Since the landfill must be serviced by a variety of solid waste collection vehicles, it is important to know whether or not there are any highway restrictions which will limit accessibility. Any restrictions would, of course, need to be eliminated in order for the development to be viable and the cost of any engineering work necessary to correct any of these problems could be said to represent an economic disincentive to choose one particular site over another.

Typical vehicles and associated height, weight and width restrictions are presented in Figure 5-5.

## FIGURE 5-5

## PACKER TRUCKS AND TRANSFER TRAILERS
## DIMENSIONS AND WEIGHT LIMITS

**PACKER TRUCKS**

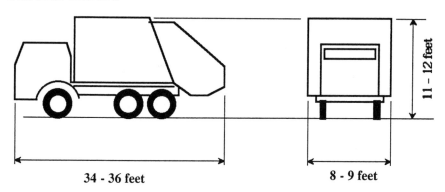

34 - 36 feet          8 - 9 feet

11 - 12 feet

**TRANSFER TRAILERS**

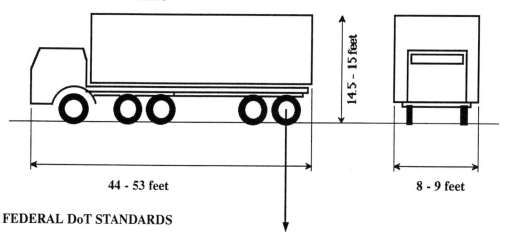

44 - 53 feet          8 - 9 feet

14.5 - 15 feet

**FEDERAL DoT STANDARDS**

MAXIMUM WEIGHT PER AXLE  =  20,000 pounds on designated roads
                            18,000 pounds on non-designated roads
MAXIMUM WEIGHT         =    80,000 pounds

**STATE DoT STANDARDS PERMIT A WIDE VARIETY OF OVERSIZE AND OVERWEGHT EXCEPTIONS TO THESE REGULATIONS**

Typical highway restrictions are shown in Table 5-2:

## TABLE 5-2
## HIGHWAY RESTRICTIONS OF CONCERN
## TO SOLID WASTE HAUL VEHICLES

Bridge height clearance
Bridge width clearance
Bridge load limit
Road width
Safe sight lines

## Traffic Impact

Areas of existing traffic congestion are often available from municipal government offices, local police, or Department of Transportation.

Landfills generate considerable traffic volume particularly during certain periods of the day. The most common high traffic periods are between 10 to 11:30 in the morning and 2 to 3 in the afternoon.

## Distance from Centroid of Waste Generation

In the overall analysis, the cost of solid waste transportation is generally at least 50% of the total cost of solid waste management. It makes sense to limit this cost as much as possible by keeping the transportation cost to a minimum.

There are elaborate computer models available which will work out the optimum location for a landfill based upon the road network, and these may be appropriate for your particular situation. In practice, the landfill siting options may be so limited that a computer model is not necessary to identify which of several options represents the least transportation cost option.

Given the locations and quantities of waste from several contributory communities, the centroid of the total waste generation in the region is very easy to establish. Once this is identified, the distance of the various sites from the centroid will establish a comparative weight for each option with the site closest to the centroid carrying the most weight for this particular criterion.

## Availability

Availability is quite simply whether or not the land is on the market. Beyond the simple device of hiring a broker, this particular criterion is often difficult to determine.

For governmental bodies, the right of eminent domain may place this particular criterion out of consideration, although this particular power is rarely exercised without misgivings. For private companies, the market forces prevail.

## Land Holding in Large Parcels

A plat book is generally the only resource necessary to plot this particular criterion. The theory behind the criterion is simply that it is preferable to negotiate with one land owner rather than a great many.

However in the real world, the chance of finding only one owner is slim, and it is more likely that some negotiation with several owners will be necessary.

## LOCAL SITING

## Preliminaries

There are six exclusionary items listed in the local siting criteria.

TABLE 5-3
EXCLUSION CRITERIA INCLUDED IN LOCAL SITING

Slope
Threatened and Endangered Species
Area of Historic Importance
Areas of Architectural Importance
Areas of Paleontological Importance
Areas of Archaeological Importance

Obtaining the necessary data to map these resources at a regional level is often extraordinarily difficult, and for that reason they are often handled in the local siting evaluation because at that level the areas to be considered have been reduced to manageable proportions.

A useful preliminary step in the local siting process is to prepare a local siting sieve map based on these exclusionary criteria only. Simply return to the modified regional siting map produced at the conclusion of Chapter 4, and reproduced here as Figure 5-6.

# FIGURE 5-6

## A TYPICAL REGIONAL SITING MAP

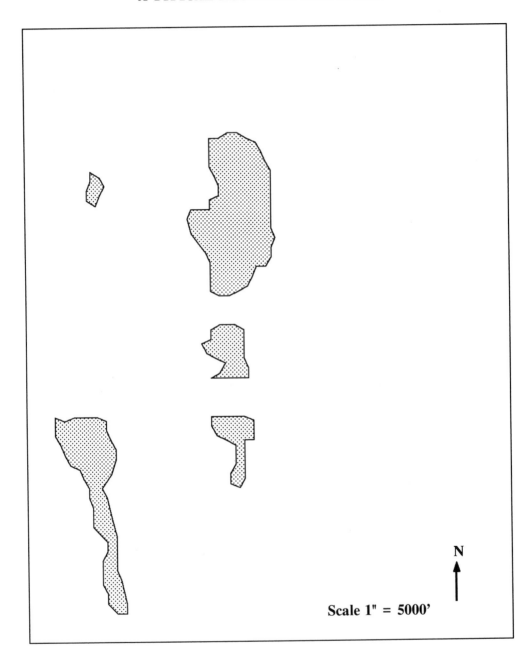

N

Scale 1" = 5000'

The object of this preliminary sieve is to map data within the shaded areas only since all other areas have been excluded. Clearly if any portions of these areas can be excluded they will reduce the time involved in the more detailed maping required for a full local study.

In order to illustrate the methodology of this preliminary sieve, let us suppose that Figure 5-7 shows the locations of an area of threatened and endangered species habitat, and a major archaeological find. Clearly these criteria eliminate a significant area from further consideration. Note also, that the remaining area is too small to comply with the minimum area criterion, and must also be excluded. This leaves just four areas to be considered in the local siting study.

# FIGURE 5-7

## A TYPICAL REGIONAL SITING MAP
## SHOWING PRELIMINARY EXCLUSION ZONES

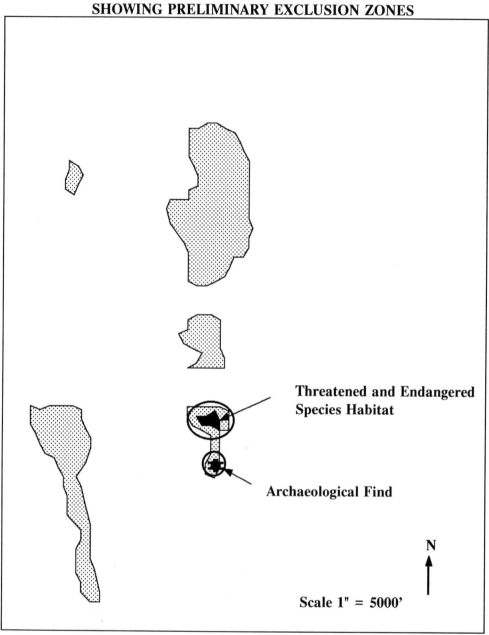

Threatened and Endangered
Species Habitat

Archaeological Find

N

Scale 1" = 5000'

These areas are typically irregular in shape, and difficult to manage in terms of future mapping, so rather than mapping irregular polygons, the areas are approximated with rectilinear boundaries as shown in Figure 5-8.

**FIGURE 5-8**

**SITING MAP**

**SHOWING AREAS FOR CONSIDERATION IN LOCAL SITING STUDY**

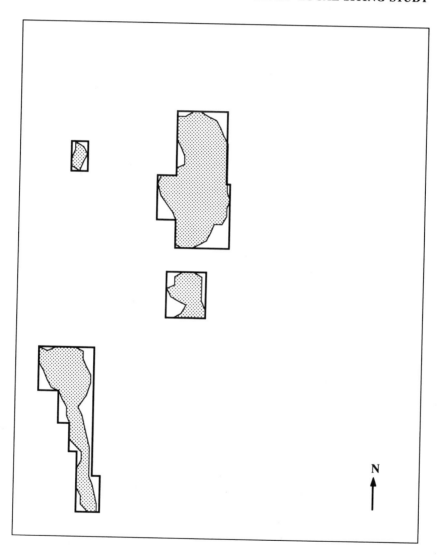

## AVAILABILITY

At this stage, one final preliminary measure is required, and that is a detailed review of land ownership patterns as they relate to the remaining study area. The ownership patterns will assist in identifying availability.

The whole question of availability is, of course, complicated by questions of price, and also ownership. A search by a governmental body is theoretically less restricted than a private company, since eminent domain may be exercised as a last resort by the governmental body. Even in this latter case, there remains a variety of existing land uses which for various reasons may render certain remaining areas off limits to landfill.

Examples of such off-limits land uses are offered here although this list is not all inclusive:

> Forest preserve areas where it is the clear policy of the governing body to deny landfill siting.

> Land held in reserve by a major corporation for the purposes of future expansion.

> Land on the route of a major highway scheduled to be developed within the next five year.

> Land held in trust by a conservation agency not previously excluded by any of the wildlife habitat criteria.

> Land currently in protracted litigation.

> Land held by a rival disposal company seeking to extend their landfill capacity.

These are all examples of real estate which has been dormant for perhaps many years for some specific purpose. In some cases, this purpose may be unannounced. In the course of eliminating large areas of real estate in the regional siting study, and focusing on the limited remaining areas in the local siting study, these hidden factors have been raised to attention. The local study team must decide on the continuing validity of these restrictions.

Often the availability of land in these circumstances may be a function of price. The forest preserve commission may be willing to forgo the use of some portion of their holdings for a limited and fixed period if the price is right. The major corporation may be willing to look elsewhere for corporate expansion if the incentive is significant.

These judgments must be made and documented, the map of candidate areas for local siting is then redrawn to reflect any changes.

The map of candidate areas for local siting may now appear as in Figure 5-8. We have assumed that the patterns of land ownership breaks down the large area at the center of our example into three, which can be conveniently considered as separate areas.

# FIGURE 5-9

## SITING MAP

## SHOWING CANDIDATE AREAS FOR DETAILED LOCAL SITING

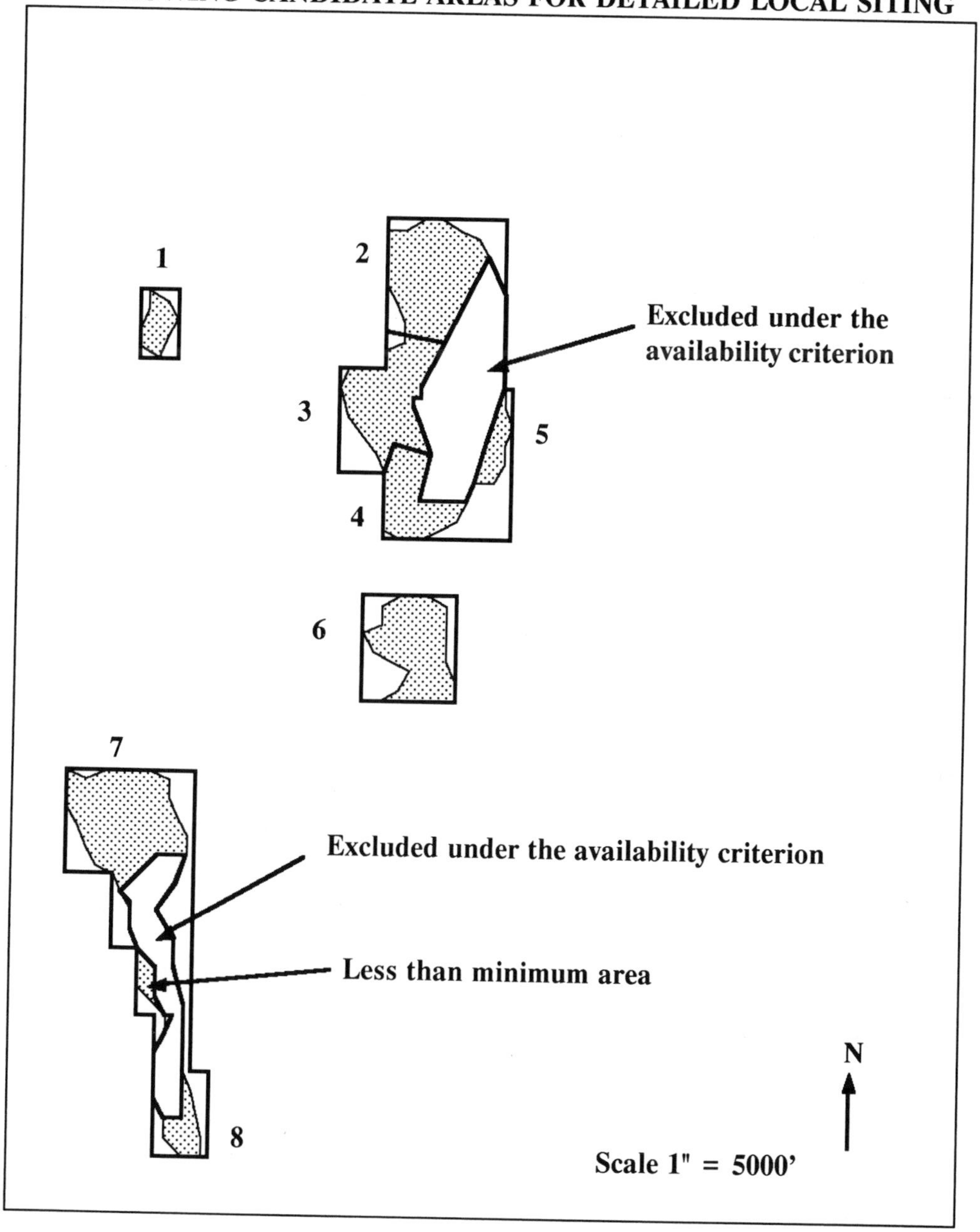

Thus the product of the regional siting study has been reduced to eight candidate areas which can now be evaluated in detail by mapping the remaining local siting criteria.

Some of these areas may be too large for the purpose, but at this stage the preferred portion of each study area is not known. As the mapping proceeds, preferred areas will develop which may center around an existing depression, or an area of limited residential well density. These preferences within candidate areas should be allowed to develop. Some areas may drop out of consideration as more detailed information is obtained. Some large areas may divide into two or more distinctly separate candidate areas. These changes should be permitted to evolve as part of the siting process until all the required information has been mapped.

# 6

# Field Verification

## INTRODUCTION

There are several items identified in the list of local criteria which require verification in the field. Up to this stage, maps have been the major source of data and this information is only as good as the latest date of revision.

Many changes in land use can take place with such rapidity that the field conditions must be constantly verified to assure that a selected site continues to be viable.

Recommended times for field verification are:

1.  When the local selection process is about to begin.
2.  Prior to making the local selection study publicly available.

## CRITERIA TO BE VERIFIED

The criteria most in need of field verification are:

> Existing Depressions
> Existing or Committed Development
> Natural Screening
> Run On Potential
> Threatened and Endangered Species
> Highway Restrictions
> Availability
> Landholding in Large Parcels

## EXISTING DEPRESSIONS

Sites which include a positive assessment for existing depression need to be checked to verify that there has been no filling which has taken place since this determination was made.

## EXISTING OR COMMITTED DEVELOPMENT

For a site to have survived the regional selection process, no existing or committed development would have been evident, so any sites being considered in the local selection process should be free of any further development. However, during the time taken to complete the regional selection process, there may have been changes which affect the status of a potential site. In addition, professionals working in this field will recognize that there is often a wide gap between what should occur in the course of land development, and what does occur. Always check local criteria.

# NATURAL SCREENING

Natural screening will serve to protect the site from overview by distinct segments of the community:

> Highway Users
> Nearby Residents
> Gathering Areas

## Highway Users

To check highway use overview, it is necessary to drive the highways in the vicinity of the site and determine the extent to which the potential site can be viewed from the highway. Office preparation involves careful review of local highway maps and identification of the major highways in the local area. Topographic maps should be used for this assessment since changes in elevation will significantly impact the number of highways from which the driver could potentially view the site. In practice, this procedure can be more difficult than it sounds unless some nearby local landmark points the way. In the absence of any local guidance, it is worth investing in several large weather balloons and tethering them at projected corner points of the proposed fill area. If the proposed final height of the landfill is known the balloons should be set at this finish level. In this way general overview and the view of the completed fill can be checked at the same time.

Remember that what is required at this stage is simply an assessment of how much natural screening is available. The fact that the balloons can be seen clearly from a particular highway does not exclude a site from further consideration. This method is merely used to compare one potential site with another based on the amount of natural screening currently available.

## Nearby Residents

A check on the availability of natural screening for nearby residents requires some detailed office preparation before embarking on the field work. All local residences should be marked on a single map. A half mile radius is suggested for this purpose unless some significant topographic feature within a mile raises a residential area to a prominent overview position in which case these areas should also be checked as special cases.

Committed future development should also be included in this survey. Depending on the definition used for the identification of this local criterion, the definition used here should be consistent. For example, where committed future development is defined as development having received plat approval, and which has been active within the last two years, this definition must be as valid at the end of the study as at the beginning.

<u>Gathering Areas</u>

Areas loosely referred to as gathering areas should also be mapped at this stage. These areas include, but are not limited to:

Schools
Churches
Community Centers
Hospitals
Recreational Areas
Shopping centers

Mapping these areas in the region of site options will help in assessing the value of natural screening.

## RUN-ON POTENTIAL

Major changes in run-on potential are unlikely, but not impossible. To avoid any future embarassment determine the up-to-date status of adjacent development before results of any local site selection process are made public. Changes likely to cause revision of any former assessment of run-on potential include:

Shopping Center Development
Parking Lot Development
Changes in Irrigation Policy
Highway Development

## THREATENED AND ENDANGERED SPECIES

The data used to determine whether or not a potential site is habitat for threatened and endangered species is probably the most subject to change of all the site selection criteria.

Unless a local wildlife study has been recently conducted, it is imperative that each site option should be evaluated in detail to determine the existence of threatened and endangered species. A wildlife expert will be required to make this assessment, and it is likely that a complete seasonal study may be required. Early spring is the season of choice for most species, but raptors and other migratory species will not appear in many areas of the country until late fall or early winter.

Determine the need for study early, and prepare for a long wait. In certain areas, the potential for threatened and endangered species habitat may be inferred, but in many areas this is by no means conclusive. The developer should be aware that the existence of threatened and endangered species has historically killed a large number of otherwise potentially good sites. The message is clear. Careful research to eliminate such sites from further consideration will save costly investment in a potential site.

## HIGHWAY RESTRICTIONS

Typically, highway restrictions exist in rural areas where non standard conditions exist, and have not been corrected in the course of development. This type of problem is unlikely to be altered by new development except that the problem may have become chronic as a result of increased traffic.

The gradual deterioration of bridge structures is a typical problem, and all potential restrictions of this kind should be checked.

## AVAILABILITY

A real estate broker is an essential element in the assessment of availability. Clearly availability is not simply a determination of whether or not the land is for sale, but an overall determination of whether or not it could be available given a favorable offer in the present market place.

## LAND HOLDING IN LARGE PARCELS

The broker who reports on availability will monitor whether or not there are changes in ownership of land in the various options.

It is worth noting that knowledge of any changes in ownership of adjacent parcels may also be valuable at a later date. Even after going through a detailed siting study, there is no guarantee that the chosen site will be approved. Final approval will depend upon a concerted public relations program, and it is as well to know the potential neighbors as early as possible in the process, and also any recent changes in ownership.

# 7

# Weighting and Rating

## INTRODUCTION

Once the local siting process has been completed, the result is a series of candidate sites as identified in Chapter 5. As we have observed in Chapter 6, the number of sites may have been modified in terms of size and availability by field verification, and other information developed during the collection of local data.

However, none of the criteria utilized in local site selection are exclusionary with the exception of those applied in the preliminaries, and the net result is a number of candidate sites, and a lot of data. This is sharply in contrast to the neat sieve of available areas which remains after the application of the regional siting criteria.

These data need to be organized in a manner that will facilitate decision making, and this can be accomplished by weighting and rating.

The methodology is essentially similar to the DRASTIC method described earlier in Chapter 2. Basically each criterion is weighted on a comparative basis, the most important item compared to others in the list being weighted as 5, and the least important weighted as 1. Once the weighting is accomplished, each candidate site can then be rated according to how it compares to the other candidates. The total score for each site is then determined according to the following formula:

$$S_A = W_1 \times R_1 + W_2 \times R_2 + W_3 \times R_3 \text{ etc.}$$

where
$S_A$ = Total score for site A
$W_1$ = Weight of first criterion
$R_1$ = Rating of first criterion

The procedure has the benefit of acknowledging that some criteria are more important than others. No matter how they are <u>rated</u> subsequently the comparative weight of each criterion will maintain balance in the selection process.

## WEIGHTING

The weight of each criterion must be determined according to its relative importance.
We will defer the question of who determines the weighting for the present. Clearly the process of assigning weights to the criteria can be delegated in many different ways. The alternatives will be discussed in detail in Chapter 8.

## TABLE 7-1

## COMPARATIVE WEIGHTS OF LOCAL SITING CRITERIA

| CRITERIA | WEIGHT |
|---|---|
| **Natural Features** | |
| Depth of Suitable Soils for Cover | 4 |
| Existing Depressions | 1 |
| Natural Screening | 2 |
| Run-on Potential | 3 |
| Residential Well Density | 5 |
| Ease of Monitoring Groundwater | 5 |
| Scenic Areas | 2 |
| Significant Depth to Groundwater Resources | 5 |
| **Land Use** | |
| Buffer Zone | 2 |
| Final Use Compatibility | 1 |
| Municipal Boundaries | 1 |
| Highway Restrictions | 3 |
| Traffic Impact | 4 |
| Distance from Centroid of Waste Generation or Transfer | 4 |
| Availability | 3 |
| Land Holding in Large Parcels | 2 |

Table 7-1 shows the none exclusionary local siting criteria, and the author's suggested weight for each criterion. It is important to emphasize that the addition of other local criteria which may be revealed during the development of the matrices will change these weights. Different geographical conditions, and different state regulatory requirements and preferences may also impact these suggested weights.

In the author's view, the process of assigning weight should be a technical activity, and rating should be either a public involvement activity, or at very least an activity delegated to a subcommittee drawn from the general public. Opinions on this subject vary considerably, and each siting committee must make a choice consistent with the special circumstances.

A review of the basis of the above weighting follows:

## Depth of Suitable Soils for Cover
Weight = 4

As noted earlier daily soil cover defines sanitary landfill practice. Its availability at reasonable depth is, therefore, extremely important to the efficient development of the landfill.

There are alternatives however. If cover soil is not available on site, it can be imported, although often at significantly increased cost. In recent years, a synthetic foaming agent has been used successfully as a daily cover material.

These alternatives impact the overall economics of the operation, but they should not impact the weighting at this stage. Indeed the ability to mitigate any of the local criteria must be ignored in the initial application of site selection criteria.

It is only later in the final determination of the best site that additional questions about mitigation must be handled.

At present, we must continue on the assumption that there is a site available among the candidates which is so ideal that questions of mitigation need not be exercised.

## Existing Depressions
Weight = 1

Existing depressions reduce the cost of site excavation to prepare air space. They are a luxury in the landfill siting business rarely encountered in practice. Useful, but not essential.

## Natural Screening
Weight = 2

Natural screening saves the cost of planted screening, and this has a useful impact on the cost of the operation.

In addition existing natural screening can be invaluable in minimizing the change in the landscape brought about by the landfill. For this reason I would judge natural screening to be of greater weight than existing depressions, but it is still in the category of useful, but not essential.

Run-on Potential
Weight = 3

This feature deserves serious study particularly since the recent passage of surface water control regulations which may ultimately require the surface water diverted by landfills to be treated before discharge.

Handling and treating rainwater run-on simply because the landfill site happens to be in the path of overland flow can be extremely costly.

Residential Well Density
Weight = 5

Since residential development within 500 feet is included in Chapter 4 as an exclusion zone, this item refers to residential development in excess of 500 feet or outlying rural property.

The potential for groundwater contamination from landfill development is a very serious concern. In a later chapter it is suggested that the impact upon water supply wells is so serious that landfill management should consider developing water quality guarantees to all well owners within a defined down gradient radius of the site.

Whether this suggestion is followed or not, it points to the seriousness of the concern, and the larger the number of wells, (or the greater the well density), the greater the overall concern.

The actual number of wells within a given radius (a half mile is suggested) should be a function of the rating process.

Ease of Monitoring Groundwater
Weight = 5

Modern landfills are required to monitor their impact upon groundwater. Clearly this is required as an early warning system, and parenthetically it should be observed that such systems must be presented in tandem with the contingency plans which lay out the actions which follow in the event that the early warning is sounded.

The early warning system itself is dependent upon a complete understanding of the groundwater regime in and around the proposed landfill.

Any complexities or uncertainties lead to the conclusion that contamination could be occurring undetected which has a depressing effect upon the acceptance level of the monitoring system.

The degree of complexity and uncertainty will have a direct bearing on the ratings. As far as weighting is concerned, the ability to monitor and to fully understand, the monitoring data must be considered of the greatest importance and the greatest weight.

## Scenic Areas
Weight = 2

The comparatively low weight assigned to scenic areas is not intended to belittle the importance of this criterion. Encroachment upon scenic areas is an important and sensitive issue.

If the life of the site is projected to be around 5 years, the significance of the impact upon a scenic area is somewhat less severe than if the projected life is to be 20 years.

The special circumstances of the scenic areas which may be potentially impacted should not be considered here, but only the relative importance compared to other factors.

## Significant Depth to Groundwater Resources
Weight = 5

No matter how well the site is designed, and no matter how many liners are included, the ultimate concern is the impact upon groundwater resources.

The separation thickness of the soils (assumed here to be of the requisite permeability by application of the regional criteria) is critical to the overall acceptability of the site.

If suggestions contained in Chapter 11 which favor water quality guarantees are included, this depth of soils to groundwater assumes an immediate economic importance.

## Buffer Zone
Weight = 2

A buffer zone is a zone of separation over which the landfill owner has control. A buffer zone may or may not include natural screening.

The important fact is that control over the buffer zone limits access, precludes further development, and allows for the possibility of remedial action in the event the groundwater monitors indicate leachate contamination has occurred.

In some states where a buffer zone is required, there is no question of evaluating a candidate site without one, and the buffer zone is simply deducted from the available land area. There is no question of excluding a buffer zone in these states. At the rating level what will be rated is the quality of the buffer zone.

In those states where a buffer zone is not a specific requirement, then the rating should reflect the absence of buffer zone by a low rating.

## Final Use Compatibility
Weight = 1

The sensible final uses for landfills are extremely limited. The post closure period is widely recognized as so critical to the overall success of the operation that few designers care to interfere with final cap design by building elaborate structures on the settling fill. Few states will permit extensive development on closed landfills.

In general, the best final uses are passive open space with fairly limited public activity, unless special design features are included to protect the final cap.

As a general rule, the question comes down to assessing the compatability of open space with whatever activity borders the site, and there are few activities which could not benefit from a little more open space.

While the rating process should be used to assess the relative merits of each alternative, the weighting process determines how important the question is in the overall scheme of things.

Final use compatability is almost always reasonably assured, and therefore, of relatively low weight.

## Municipal Boundaries
Weight = 1

For the landfill to be contained within municipal boundaries is largely a function of control. While this may be an important factor for some municipalities, for others it is equally important that it be outside the boundary for reasons of liability and identification.

This criterion is of low priority in terms of its impact upon the operation of the facility, and the proximity of the site to the municipal boundary rarely has any impact on the design. For these reasons, in the author's view, this criterion carries a low weight.

## Highway Restrictions
Weight = 3

The presence of any highway restrictions on any of the routes to and from the landfill can be serious. Remember that the degree of seriousness is a function of rating.

What is being weighted here is simply how concerns about highway restrictions compare to concerns about depth to groundwater at the top of the scale, or existing depressions or final use compatibility at the bottom of the scale.

## Traffic Impact
Weight = 4

In the author's view, traffic impact is somewhat more significant than highway restrictions.

While highway restrictions may be a nuisance in terms of access to the facility, it will be a nuisance restricted to a limited number of people, and possibly only the operator.

Traffic impact, on the other hand, will generally create a nuisance to a large number of people, and cast a shadow over the whole operation.

## Distance from Center of Waste Generation or Transfer
Weight = 4

The overall cost of the transportation system is closely tied to this criterion. Long haul to distant landfills increases energy and maintenance costs. In addition, long haul takes collection vehicles out of the collection mode necessitating higher costs for replacement vehicles.

## Availability
Weight = 3

The question is not simply whether or not the site is on the market, but whether current owners would have an interest in selling if the price was right. The actual input data for the local mapping of this criterion should be provided by a qualified real estate assessor. The data is typically in the form of a judgement regarding availability at the right price, and may be conveniently presented in the categories: unlikely, likely, and highly likely. As we will see later, these categories can be easily represented in the rating process.

The question here is how much importance should be placed on this criterion compared to others in the criteria list.

## Land Holding in Large Parcels
Weight = 2

It is clearly easier and more convenient to negotiate land purchase with the owner of a single large parcel of land, than it would be to negotiate with a great many people, each of whom owns a small piece of the projected site.

## RATING

The rating of each criterion depends entirely upon the local conditions which prevail at each candidate site.

Ratings vary considerably depending upon local preferences.

Some general guidelines are offered as a means of directing raters toward the elements which need to be considered.

Depth of Suitable Soils for Cover

The categories to be rated are:

| Rating Categories | Suggested Rating |
|---|---|
| Soils at reasonable depth (0 - 20 feet) | 6 - 10 |
| Soils at great depth (20 - 50 feet) | 3 - 6 |
| Need to import (minimal soil available) | 1 - 3 |
| Need to use synthetic cover (minimal soil available) | 0 - 1 |

Factors which should be considered in rating these categories are:

| | | |
|---|---|---|
| Overall cost of additional excavation to obtain cover necessary. | - | Excavation costs increase as depth of excavation increases. Beyond 20 feet, double handling may be |
| Impact of stockpiling on the efficiency of the operation | - | Where cover soil is available at great depth, significant stockpiling is necessary. If the site area is limited, this will have a significant impact upon the efficiency of the operation. |
| Cost impact of importation | - | The import rating can be increased if available sources of suitable cover are readily available. |
| Cost impact of synthetic cover | | Synthetic cover material is more costly per square yard. Successful use depends upon preparation of each daily cell beyond that needed for soil cover. |
| Visual impact of synthetic cover | - | Synthetic cover materials are easily distinguished at a distance. |

## Existing Depressions

Categories to be rated are:

| Rating Categories | Suggested Rating |
|---|---|
| Existing depression(s) provide greater than 25% of air space | 5 - 10 |
| greater than 10% of air space | 3 - 5 |
| greater than 5% of air space | 1 - 3 |
| greater than 2% of air space | 0 - 1 |

The single factor considered in rating these categories is:

Overall savings in excavation costs.

## Natural Screening

Categories to be rated are:

| Rating Categories | Suggested Rating |
|---|---|
| Screens greater than 25% of perimeter of landfill | 5 - 10 |
| greater than 10% | 3 - 5 |
| greater than 5% | 1 - 3 |
| greater than 2% | 0 - 1 |

Factors which are considered in rating these categories are:

| | | |
|---|---|---|
| Density of screening | - | The ideal screen is impervious to seasonal change, i.e. solid objects or coniferous woodland. The woodland screen should be rated down for deciduous species. |
| Jurisdiction over screening | - | A natural woodland screen should be highly rated if it is owned by the operator or protected under a binding regulation such as a forest preserve. |

## Run-On Potential

Categories to be rated are:

| Rating Categories | Suggested Rating |
|---|---|
| In the top 25% of watershed | 8 - 10 |
| In the top 50% of watershed | 6 - 8 |
| In the top 75% of watershed | 4 - 6 |
| In the bottom 25% of watershed | 0 - 4 |

Factors which are considered in rating these categories are:

| | | |
|---|---|---|
| Local topography within watershed which may reduce run-on potential | - | Areas of high ground within the watershed may have reduced run-on potential |
| Limited rainfall | - | Areas where net infiltration is low, may choose to discount this category. |

## Residential Well Density

Categories to be rated are:

| Rating Categories | Suggested Rating |
|---|---|
| Less than 5 wells within a half mile of the periphery of the landfill footprint | 8 - 10 |
| Less than 10 | 6 - 8 |
| Less than 15 | 4 - 6 |
| Greater than 20 | 0 - 4 |

Factors which are considered in rating these categories are:

| | | |
|---|---|---|
| Numbers of households connected to each well | - | Residential wells with multiple users should be rated greater than single use wells |
| Down gradient wells | - | Wells down gradient of the landfill are more vulnerable to contamination than those located up gradient. |

## Ease of Monitoring Groundwater

Categories to be rated are:

| Rating Categories | Suggested Rating |
|---|---|
| Straightforward, easily monitored groundwater | 5 - 10 |
| Complex monitoring conditions due to hydrogeology | 2 - 5 |
| Complex monitoring conditions due to existing sources of groundwater contamination | 0 - 2 |

Factors which are considered in rating these categories are:

| | | |
|---|---|---|
| Degree of complexity hydrogeology | - | Ratings should reflect the ability to monitor even of under complex conditions. |
| Extent of existing contamination | - | Where existing contamination is widespread and pervasive, ratings should be reduced. |
| Similarity of contamination | - | Conditions in which existing contamination includes common constituents of landfill leachate should be down-rated. |

## Scenic Areas

Categories to be rated are:

| Rating Categories | Suggested Rating |
|---|---|
| No scenic areas impacted | 7 - 10 |
| Scenic areas of local interest impacted | 4 - 7 |
| Scenic areas of regional interest | 2 - 4 |
| Scenic areas of national interest | 0 - 2 |

Factors which are considered in rating these categories are:

| | | |
|---|---|---|
| The extent of local interest in the impacted scenic area | - | Areas of limited local interest are rated lower |
| The projected life of the site | - | All impacts on scenic areas for a limited period the should be up-rated. |
| The configuration of final use | - | If the finish topography of landfill will not be sharply in contrast with that of the local area, the overall impact on the scenic area should be up-rated. |

## Significant Depth to Groundwater Resources

Categories to be rated are:

| Rating Categories | Suggested Rating |
|---|---|
| No grouhdwater resources underlying the landfill and witbin a half file of the landfill footprint. | 8 - 10 |
| No groundwater resources underlying the landfill footprint | 6 - 8 |
| Depth to underlying groundwater resources in excess of 80 feet | 4 - 6 |
| Depth to underlying groundwater resources in excess of 50 feet | 0 - 4 |

Factors which are considered in rating these categories are:

| | | |
|---|---|---|
| Quality of groundwater resources | - | In regions where reliable groundwater information exists, some up-rating may be considered for non-potable resources. |
| Limited extent of resource | - | For confined aquifers where useage is limited groundwater to less than 5 residential wells, some up-grading may be considered. |

## Buffer Zone

Categories to be rated are:

| Rating Categories | Suggested Rating |
|---|---|
| Buffer zone on all sides greater than 150 feet | 9 - 10 |
| Buffer zone on all sides greater than 75 feet | 8 - 9 |
| Buffer zone over 75% of perimeter of 75 feet or more | 5 - 7 |
| Buffer zone over 50% of perimeter of 75 feet or more | 4 - 5 |
| Buffer zone over 25% of perimeter of 75 feet or more | 3 - 4 |
| Buffer zone over less than 25% of perimeter | 0 - 3 |

Factors which are considered in rating these categories are:

| | | |
|---|---|---|
| Landfill owner control over buffer zone | - | Buffer zone over which the landfill owner has full control is more highly rated. |
| Remote location | - | In remote locations the buffer zone category may be omitted. |

<u>Final Use Capability</u>

Categories to be rated are:

| Rating Categories | Suggested Rating |
|---|---|
| Proposed final use fills a clear need in the local area. | 7 - 10 |
| Final use duplicates existing resources, but it is not incompatible with the area. | 3 - 6 |
| Final use is incompatible with the area | 0 - 3 |

Factors which are considered in rating these categories are:

| Linkage with existing open space | - | Newly created open space which provides linkage between existing areas of open space should be up-rated. |
|---|---|---|
| Scarcity of open space | - | Open space which fills a regional need should be up-rated. |

<u>Municipal Boundaries</u>

Categories to be rated are:

| Rating Categories | Suggested Rating |
|---|---|
| Within the municipal boundary | 8 - 10 |
| Within the area of influence of the municipality | 5 - 7 |
| Outside the municipal area of influence, but in an area where county control is strong | 2 - 4 |
| Outside the municipal area of influence, but in an area where county control is weak | 0 - 2 |

Factors which are considered in rating these categories are:

Local control            -       In the author's view, local control is always preferable to regional or state control, because it is the only authority on scene on a daily basis. However, this authority is worthless if the financial resources are not available to police the operation. Lack of local resources should, therefore, be down-rated in these circumstances.

## Highway Restrictions

Categories to be rated are:

| Rating Categories | Suggested Rating |
|---|:---:|
| No restrictions to access | 9 - 10 |
| Limited restriction to access to over 50% of the routes to the landfill | 8 - 9 |
| Limited restriction to access from all directions | 7 - 8 |
| Serious restriction to access to over 50% of the routes to the landfill | 4 - 6 |
| Serious restriction to access from all directions | 0 - 3 |

## Traffic Impact

Categories to be rated are:

| Rating Categories | Suggested Rating |
|---|:---:|
| No traffic impact | 8 - 10 |
| Limited traffic impact restricted to the area of the landfill | 6 - 8 |
| Limited traffic impact on several access routes | 4 - 5 |
| Average traffic impact in local area | 2 - 4 |
| Significant traffic impact in the local area | 0 - 2 |

Factors which are considered in rating these categories are:

Timing          -      As noted earlier, the times when traffic volume is at its height are 10 - 11:30 A.M., and 2 - 3 P.M. which are not typically periods of high traffic volume. This should be considered in rating.

Gathering areas    -      Gathering areas such as nursing homes, day care centers, shopping centers, etc. would be impacted by the off peak times of typical landfill traffic. These conditions should be down-rated.

## Distance From Center of Waste Generation

Categories to be rated are:

| Rating Categories | Suggested Rating |
| --- | --- |
| Within 10 miles | 8 - 10 |
| Within 20 miles | 7 - 8 |
| Within 30 miles | 5 - 6 |
| Within 40 miles | 3 - 4 |
| Within 50 miles | 1 - 2 |
| In excess of 50 miles | 0 - 1 |

Factors which are considered in rating these categories are:

Major highways    -      Where major highways are limited, the mileage should be determined as road miles. For areas with good coverage by major highways "crow fly" miles are adequate for this assessment unless the final scores are very close.

Toll roads        -      As a general rule, sold waste vehicles do not use toll roads.

## Availability

Categories to be rated are:

| Rating Categories | Suggested Rating |
|---|---|
| Highly like to sell | 8 - 10 |
| Likely to sell | 5 - 7 |
| Unlikely to sell | 2 - 4 |
| Definitely no sale | 0 - 1 |

Factors which are considered in rating these categories are:

Price          -          Price should be omitted from consideration during this phase. The whole question of cost as a factor in siting is discussed at the conclusion of this chapter.

## Land Holding in Large Parcels

Categories to be rated are:

| Rating Categories | Suggested Rating |
|---|---|
| One owner | 9 - 10 |
| Two owners | 7 - 8 |
| Three owners | 5 - 6 |
| Three to five owners | 3 - 5 |
| Five to ten owners | 2 - 3 |
| More than ten | 0 - 1 |

Factors which are considered in rating these categories are:

Affiliation               -         Land owned by several owners all of whom are in partnership or affiliated in same manner should be up-rated.

## THE QUESTION OF COST

It should be apparent at this stage that many of the regional and local criteria are cost related. While the search for ideal siting criteria is in essence a search for areas of minimal environmental impact, this can also be translated as a search for the low cost solution. The avoidance of such areas as wetlands, flood plains, surface waters, groundwater, fault zones, seismic zones, unstable areas, expansive soils, and subsidence zones reduces the high costs of mitigating such difficult conditions, and that points toward the low cost solution.

Placing a protection zone around municipal wells reduces the potential of environmental liability. Increasing proximity of a landfill to major highways reduces the high cost of building access roads. All of these provisions favor the low cost solution.

Similarly, many of the local criteria also translate into efforts to reduce overall costs.

The search for suitable soils at reasonable depth for cover material, existing depressions, natural screening, ease of monitoring, significant depth to groundwater, and reasonable distance from centroid of waste generation, all lead to low costs for landfill development as well as reduced environmental impact.

Similarly, the avoidance of areas with high run-on potential, significant slope, highway restrictions, and traffic impacts, leads to significant cost reductions.

Thus, once the final scores are tabulated, and the highest scoring site is identified, the chosen site will represent a combination of low cost solution, and minimal environmental impact.

The temptation to manipulate the final scores may be strong at this stage, particularly if the difference in the scores of the top two or three candidates is very small. This temptation should be avoided because it is essentially meaningless. Planting trees to raise the screening score, or adding land to raise the buffer zone score has no real meaning when the majority of the siting criteria themselves incorporate a significant measure of cost reduction.

The only way in which a true cost comparison could be made would be to complete preliminary designs for landfills at the top two or three candidate sites, prepare detailed estimates, and compare costs. However, this approach is not recommended because it cuts across the essential assumption of the siting process, which is that the best site is the one which combines the lowest practicable cost with the least environmental impact.

A detailed cost comparison of candidate sites is strongly discouraged since it will inevitably focus attention on the single feature of cost while obscuring the goal of minimizing environmental impact.

None of the foregoing should be taken to suggest that further mitigation is not recommended. Once a site is selected using the selection criteria described, every effort should be made to mitigate. This should not be done for the purpose of raising the score, but for the purpose of enhancing the final design.

# 8

# Public Participation and the Siting Process

## INTRODUCTION

Public participation in the siting process is a necessity because, ultimately, whether or not the final plan works depends upon public approval. This need for public participation may not be immediately obvious at first glance since solid waste management is typically delegated to the public official or to private enterprise.

However, delegation in this area is only temporary. In the final analysis, it is the public which makes the final decision.

## A TWO PART PROCESS

In practice, the process of siting falls into two distinct parts. The first is regional siting, and the second is local siting.

### Regional Siting

Regional siting typically falls to the engineer (or planner) to struggle with, because general public participation is not easily engaged until local sites are identified. Also, the issues on a regional scale are often considered to be too broad and general to generate wide public interest.

Nevertheless, every effort should still be made to obtain some measure of public involvement. A series of region-wide meetings should be held.

The principal purpose of these meetings is to introduce the public to the whole general approach to solid waste management planning, and to engage public participation in the siting process.

Where siting is undertaken by private enterprise, there appears to be widespread reluctance to engage public involvement at any level until a final decision is made. This attitude is often justified on the grounds that public involvement is generally negative, and even hostile to the siting process.

I believe it would be more correct to say that the public is often hostile to the site, not the process. In most situations where the site is presented as a given, the process is a closely guarded secret, and it is hard to tell whether the public is hostile to the site alone, or their lack of involvement in the siting process. In fact, the general public is usually very sympathetic to the siting process. It is only when a site is chosen that the public is polarized, and that section of the community most affected by the proposal becomes vocal.

In the author's opinion when the public becomes vocal about a site, it is too late to start the public involvement process.

Whether siting is public or private, the public must be involved from the beginning. Public involvement is the only way to build the consensus of support needed to get beyond the inevitable vocal stage which occurs when the site is selected.

The regional siting process is the time to get the public involved. They should be involved in setting the overall guidelines, and sizing the site.

As a practical matter it may be appropriate to appoint a subcommittee, but at least one region-wide solid waste management meeting is a must in order to get the ball rolling, and to oversee the selection of the subcommittee.

<u>Local Siting</u>

Local siting is the next step in the planning process. Interest in siting is typically directly proportional to how accurately each site is identified. This is unfortunate, because it would be preferable, from a public involvement perspective, to engage the interested public in the siting process before the site is selected rather than afterwards. In this way, the public could be proactive rather than reactive. We need some new strategies.

PROBLEM DEFINITION MEETINGS

The earliest meetings on the subject of solid waste management should be problem definition meetings. At these early meetings, the agenda should be as follows:

TABLE 8-1

AGENDA FOR STAGE 1

PUBLIC INVOLVEMENT

1. Introductions

2. Purpose of Meeting

3. Problem Definition

    a. The rising solid waste generation rate
    b. The increased cost of disposal
    c. The increased distance of haul
    d. The air emissions associated with the long haul
    e. The remaining life of the disposal facility

4. Management Strategy

    a. The proposed level of recycle
    b. The alternative disposal techniques evaluated
    c. The need for landfill space
    d. The land area required
    e. The general methodology of site selection

5. Wrap up Comments

A subcommittee should be appointed at one of these meetings in order to advise the planning team of the public interest. This subcommittee should be involved in identifying input to the various matrices used to identify site selection criteria.

## REGIONAL SITING MEETINGS

Once the planning team has completed the regional site selection process, the regional maps are an ideal focus for a second round of meetings.

At this stage, there should be more public involvement, since there will be a general interest in where the landfill will be located. Since the regional siting map is not site specific, there is little risk of generating massive public opposition.

The agenda for this round of meetings should be as follows:

TABLE 8-2

AGENDA FOR STAGE 2

PUBLIC INVOLVEMENT

1. Introduction

2. Thanks to Members of Subcommittee

3. Purpose of Meeting

4. Review of Regional Site Selection Map

5. The Next Step

    a. Review of Local Site Selection Criteria

    b. The Importance of Weighting and Rating

6. Appointment of Subcommittee

7. Wrap Up Comments

Once a site has been selected, the standard approach is to hold a public meeting. It appears that some public blessing is believed to be bestowed upon a selection by this process. However, in practice the public blessing is often one which consigns the choice to purgatory.

In general public meetings held to confirm a selection do not work. Whether it is easier at such meetings to express dissent rather than agreement, or whether such meetings attract only committed dissenters, it is difficult to determine, but generally speaking public meetings held after a site has been selected rarely work out to be the straightforward forum they are reported to be.

The plain fact is that every resident has a different perspective on the advantages and disadvantages of a particular selection. There is not enough time in the duration of a public meeting to meet these different perspectives face to face, and review each one, particularly when a verbal battle is taking place with the most outspoken opponents of the selection.

## Shoe Leather Public Involvement

What is needed is shoe leather public involvement. A group capable of presenting the selected site in all its various ramifications must be chosen, and they must be willing to meet one-on-one with members of the community within one mile of the proposed facility. To dispel any misunderstanding, it should be clear that what is proposed here is door to door diplomacy. Everyone within at least one mile of the facility should have a separate hearing.

Typically, these representatives of the selection committee are selected from the advisory committee who conducted the site selection process. They should be trained by the site selection consultant or advisor, and they should be prepared with some simple audio visual aids to get the point across.

The best equipment for those embarking on this activity is the ability to be a good listener. The most important single goal for this kind of public involvement is that members of the public should know that they are being heard. This is not the same as being sure that all the questions are answered!

The representative of the advisory committee should be very careful to avoid the trap of answering every question and assuming from this that the member of the public concerned has been heard. In many cases, all this means is that the representative of the advisory committee has been heard!

There are three simple tools which the representative needs to take to these meetings to measure success. They are as follows:

1. The water jug analogy.
2. A willingness to not know.
3. Good graphics.

## The Water Jug Analogy

Representatives must be trained to understand that each member of the public has valid and reasonable concerns which need to be expressed. If the representative can understand that these concerns are like water in a water jug, it can be an extremely useful analogy.

The representatives function is to see that John Q Public has the opportunity to get out all his concerns. Empty the water jug. There are various phrases which can be used to determine whether or not this point has been reached:

1. I just want to be sure I have heard all your concerns.
2. Can you think of any other points we haven't covered?
3. If you think of anything else, will you be sure to get back to me?

The whole point of this activity is to overcome the simple fact that most members of the general public feel that their concerns are very rarely <u>heard.</u> True, they are frequently given political choices, but it should be very clear that is not the same as being heard. If John Q Rep can achieve nothing else in this meeting except to convince John Q Public that he/she is being heard, that alone is a significant step forward.

## A Willingness To Not Know

All too often those in possession of information (in this case, facts on why the site was chosen) are all too anxious to provide the answers. Typically John Q Rep leaves the meeting feeling that it has been a great success, and John Q Public is left with a feeling that he was the recipient of a canned speech.

John Q Rep's ego should be parked at the door for these meetings. At the end of the water jug analogy appropriate comments are as follows:

1. I made a list of your concerns while you were talking. If I read them over, would you tell me if I missed anything?

2. Let me see if I can answer some of the questions you have raised.

3. I would like to try to answer some of your concerns, but don't hesitate to stop me if another question comes to mind.

At the end of the responses to concerns, John Q Rep should be willing to say:

"I don't have enough information to answer the question about X, but I will definitely get back to you."

In other words, be willing to not know when faced with a new question. It is far better for John Q Rep to admit he/she does not know, and promise to return with an answer.

There are many benefits to this approach. The questioner feels his concerns are being taken seriously; he asked a good question because more research is needed to answer it, and finally when John Q Rep returns with the answer, there is a new opportunity to be sure that the water jug of concerns is truly empty.

Good Graphics

Good graphics are essential because they add emphasis to the spoken word. Sometimes they are a necessary adjunct to what is being said, and they make it easier to understand difficult concepts.

However, good graphics can be too good. In this context slides are definitely out. Also out are slick card displays, or multi-colored presentations. Ideally, the graphics should be those used in the site selection report or simplified versions of the same prepared as an introduction to graphics used in the report.

If John Q Rep has any facility as an artist, simple line drawings prepared at the meeting are best because they directly address the question raised and they are clearly not canned.

<u>Local Meetings</u>

After these face to face contacts have been made, it is essential to hold at least one local meeting. This meeting should be as close to the site as you can find a reasonable meeting place, and it should be primarily a review of the questions, answers and concerns raised during the weeks of one-on-one contact.

This meeting is important to confirm that one-on-one meetings were not intended to divide the community, nor were different people told different things.

The agenda at this stage should be roughly along the following lines:

TABLE 8-3

AGENDA FOR STAGE 3

PUBLIC INVOLVEMENT

---

```
1.   Introduction

2.   Purpose of Meeting

3.   Summary of Contacts made since the last
     meeting, and reason for these one-on-one
     meetings.

4.   Questions most frequently raised, and
     answers given.

5.   Agreements to be made a part of the plan
     as a result of meetings.

6.   Plans for further meetings.

7.   Wrap Up Comments.
```

## PUBLIC PARTICIPATION OPTIONS

A proactive local siting process envisages public participation in the selection of siting criteria, and weighting and/or rating of local criteria.  As a practical matter, this level of public participation cannot take place in a vacuum, and some level of general introduction to the type of factors typically included is necessary.  One use of this book may be to provide the criteria identified as a starting point.  However, once this general introduction is completed, the public should be free to propose criteria to be included in the selection process.

It is a matter of choice as to how the public might participate in this process. The options for consideration are as follows:

Option 1   Criteria selection, weighting, and rating by a siting committee

Option 2   Criteria selection, weighting, and rating by a siting committee and by a siting subcommittee appointed by the public at an open forum

Option 3   Criteria selection, weighting, and rating by a siting committee and the general public at an open forum

A second set of options entails the separation of the criteria selection process from the weighting and rating process. The basis for this separation is that the selection of siting criteria involves somewhat more technical knowledge and familiarity with the alternatives available. The weighting and rating process involves a general assessment of local "feelings" on various aspects of siting. These options are:

Option 4   Criteria selection by a siting committee and weighting and rating by a weighting and rating subcommittee appointed by a public forum for that purpose.

Option 5   Criteria selection by a siting committee and weighting and rating by the general public at an open forum

A third set of options entails the separation of criteria selection and weighting from the rating process. The basis for this separation is that criteria selection _and_ weighting involve a certain level of technical knowledge, and the rating process can be considered an appropriate level on which to involve the public. These options are:

Option 6   Criteria selection and weighting by a siting committee, and rating by a rating subcommittee appointed by a public forum for that purpose.

Option 7   Criteria selection and weighting by a siting subcommittee, and rating by the general public at an open forum.

To summarize, there are three critical steps remaining in the next stage of the siting process - criteria selection, weighting, and rating. There are three possible groups of people involved in making these selections - a siting committee, a publicly appointed siting subcommittee or a public forum.

The following table summarizes the choices:

TABLE 8-4
PUBLIC PARTICIPATION OPTIONS
Principal Tasks

| OPTIONS | CRITERIA SELECTION | WEIGHTING | RATING |
|---|---|---|---|
| 1 | Siting Committee Only | Siting Committee Only | Siting Committee Only |
| 2. | Siting Committee + Siting Subcommittee | Siting Committee + Siting Subcommittee | Siting Committee + Siting Subcommittee |
| 3. | Siting Committee + Public Forum | Siting Committee + Public Forum | Siting Committee + Public Forum |
| 4. | Siting Committee | Weighting and Rating Subcommittee | Weighting and Rating Subcommittee |
| 5. | Siting Committee | Public Forum | Public Forum |
| 6. | Siting Committee Only | Siting Committee Only | Rating Subcommittee |
| 7 | Siting Committee Only | Siting Committee Only | Public Forum |

There are advantages and disadvantages to each of these options.

Option 1 - The Siting Committee Only

A siting committee is a body appointed to carry out activities pursuant to siting. This body is most familiar with the various criteria and most likely to understand the various implications of the weighting process.

From a practical viewpoint, the time necessary to identify criteria, weights and ratings will be minimal under this option compared to any other option. However, the committee is a small group and not necessarily representative of citizens at large.

Option 2 - Siting Committee + Siting Subcommittee

A siting subcommittee is easy to manage. After a short briefing, the subcommittee could function very efficiently, almost as an extension of the siting committee. This has the advantage of involving a larger number of people in reaching the decisions which will lead to siting. However, members of committees like this are frequently discounted if the results are unpopular with the general public.

<u>Option 3 - Siting Committee + Public Forum</u>

Including a public forum in the decision making process involves the most people in the collective responsibility. This approach also has the advantage of being the most democratic and, potentially, the most fun.

As far as weighting and rating are concerned, this approach could work well and be interesting and stimulating. Choosing siting criteria could be more difficult, since this will involve reaching a consensus which is not always an easy task in public meetings.

<u>Option 4 - Criteria Selection by Siting Committee/Weighting and Rating by Siting Committee + Siting Subcommittee</u>

This approach is rational and manageable with little prospect of error or difficulty. The disadvantage is that the subcommittee may not have very much standing in the communities who are later affected by the results.

<u>Option 5 - Criteria Selection by Siting Committee/Weighting and Rating by Siting Committee + Public Forum</u>

This approach is also rational. The public forum could be workable, because everyone would have a vote on weighting and rating.

The disadvantage is that the public forum may not be satisfied with weighting and rating criteria they did not select, and there may be concern over what might be perceived as a tame and minor role.

<u>Option 6 - Criteria Selection and Weighting by Siting Committee and Rating by a Rating Subcommittee.</u>

Again this approach is rational from the point of view that the siting committee is the most knowledgeable, and criteria selection and weighting do require a certain amount of technical background. A rating subcommittee could then develop the rating system.

The disadvantage is that this may appear to be a "managed" process with a resultant lack of respect for the final choice.

<u>Option 7 - Criteria Selection and Weighting by Siting Committee/Rating by Public Forum.</u>

The advantages of this option are the same as Option 6 plus the fact that the public forum is an ideal vehicle for rating. The weighting process itself depends upon a certain level of technical knowledge while rating is really an expression of local feelings.

## SUMMARY

Public participation in the siting process is strongly recommended. Frequently, the problem is how to involve the public effectively. Specific tasks such as criteria selection weighting or rating provide an effective and stimulating vehicle for some active participation. The only question is practicality.

As a believer in the underlying good sense of the general public, the writer favors option 3 in the belief that there is no deadlock that cannot be resolved. The next best options would be option 5 or 7.

## AFTER WEIGHTING AND RATING

When the weighting and rating activity is concluded, the site selection process is comparatively simple. Those sites ranking highest in aggregate weight and rating are the first choice for each selected technology, and the remainder ranked in order of preference according to their respective weights and ratings.

The important aspect of public participation to be noted at this stage is that once sites are identified, public interest is significantly altered. To a considerable extent, the specific members of the public who were intimately involved up to this point are overshadowed by the often more vociferous opponents of each specific siting decision. The support of members of the public who have stayed with the process from the beginning is very important during this period, but it is difficult to sustain in the face of the more partisan concerns of those immediately affected by a siting decision.

There appears to be no antidote to this condition. The process must simply go forward sustained to a large extent by the fact that the siting procedure itself is sound, fair and defensible.

As far as public participation is concerned, this is the point at which the process moves from the academic reaches of regional planning to the trenches of local planning. The venue of the public forum shifts to a meeting place as close to the selected site as possible. The content of the meeting will be dictated by the technology in question, but whatever the technology, it cannot be emphasized enough that the presenter should Have Good Graphics. The graphic presentation should be clear, concise and straightforward without a great deal of technical jargon.

# Case Study
# Lake County, Illinois

## INTRODUCTION

During 1988 - 1989, one of the most populous counties outside the City of Chicago contracted with the author to perform a regional site selection study for several types of regional pollution control facilities.

Lake County, Illinois covers a total area of 454 square miles, and lies within a range of 26 to 53 miles of downtown Chicago on the north side of the City.

The County has generously agreed to the use of the final maps to illustrate how the method described in these pages is used in practice. It is important to note that what is described here is the Regional Plan, and a later step to be conducted at a county level remains to be done.

It is also important to note that grant-in-aid was available to the county in the form of services from the Computer Division of the Illinois Department of Energy and Natural Resources.

Since this aid was available on a one time basis, it was decided to include as much data as possible into the regional selection process. Several criteria which are normally considered as local selection criteria were included in the regional criteria in order to make the best use of the available aid.

## INTRODUCTION TO THE LAKE COUNTY
## SOLID WASTE MANAGEMENT PLAN

Those seeking in these pages for the exact geographical location of proposed regional solid waste facilities will be disappointed. It is not the intention of the Joint Action Solid Waste Planning Agency to identify specific sites at this stage in the planning process.

The most which can be accomplished at this stage is to identify those areas which are most suitable for such facilities. This is generally referred to as Phase I of the siting process or Regional Site Selection. At a later stage, beyond the scope of this project, additional siting criteria must be identified in order to weigh the degree of suitability of each site and to make a preliminary site selection.

Even after preliminary site selection, the process is not complete since information must be field checked to determine the accuracy of the regional data. This is the time when options may be purchased and preliminary borings are made. Since this is an expensive procedure, the true value of the siting process will be apparent in that it narrows the options which need to be evaluated in detail.

Phase I involves the identification of siting criteria which represent the most important basic requirements for a good facility. In many cases these criteria are so basic that they are part of the regulations. A good example of this kind of criterion is the prohibition against siting in flood plains. Floodplains are identified as an exclusion zone for all types of solid waste management technology.

For sanitary landfills the availability of suitable soils is a basic requirement. Areas lacking suitable soils are, therefore, mapped as exclusion zones. This applies only to landfills and composting facilities since waste-to-energy systems and transfer stations have no requirement for suitable soils.

For general guidance these criteria are identified under five different categories:

> Natural Features
> Cultural/Institutional Resources
> Development
> Highways
> Additional Features

Regional criteria are discussed in the following pages and their relevance to each potential solid waste technology is also identified.

Each of the criteria is mapped as an exclusion zone. That is to say the characteristics which need to be left out of further consideration are plotted as areas to be excluded in the final analysis. All the maps of exclusion zones are then added together and the result is a single composite map for each technology option which identifies the areas most suitable for further investigation.


## NATURAL FEATURES

Floodplains

In order to limit the potential for inundation of solid waste management facilities by floodwaters, state and federal regulations restrict construction in the floodplain where the construction has no special flood protection. Since flood protection measures are costly for landfills, the floodplain is usually identified as an exclusion zone.

In Lake County the one hundred year floodplain is not mapped for the entire county. For convenience the flood-of-record map was used. In some cases a flood of greater magnitude than a 100 year flood is represented and in some cases less. For regional siting purposes this map was considered adequate. (A one hundred year flood is one which has the probability of being equalled or exceeded once every hundred years).

## Wetlands

Wetlands are typically excluded from consideration as sites for solid waste management facilities because wetland areas are considered a valuable ecological resource. They provide habitat for a wide variety of species of flora and fauna. Wetlands also present significant design problems for most solid waste activities. A special permit is required from the U.S. Army Corps of Engineers to develop in wetland areas.

Wetland areas have been mapped as exclusion zones for all solid waste management technologies. The wetlands map utilized was provided by the Illinois State Department of Conservation and is identified as The National Wetlands inventory.

## Threatened and Endangered Species

Areas which provide habitat for threatened and endangered species are protected by several different statutes. Solid waste management facilities are specifically excluded from these areas.

However, the mapping of these areas presents a problem because only reported sighting locations are recorded. Thus only point locations were available in map form as opposed to specific areas. These sighting locations were therefore supplemented by data from a natural areas map on the assumption that these natural areas would identify the habitat area in terms of general range. In some cases no natural area could be found to identify the area, so the sighting locations were identified with the view that in a later phase the criterion of threatened and endangered species would be checked again in more detail.

Another problem with threatened and endangered species records is that they only report what has been sighted. There may be other such species in Lake County, but in the absence of an official sighting they have yet to be recognized. Clearly, this is another reason to review the findings when more specific sites are identified. The sighting locations were provided by the Illinois State Natural History Survey. The Natural Areas Inventory was provided by the Illinois Department of Conservation.

## Suitable Soils

For sanitary landfills, perhaps the single most important criterion is the availability of suitable soils. Suitable soils for sanitary landfills in northern Illinois are defined as the silty or clayey tills with a permeability in the range $1\times10^{-7}$ to $1\times10^{-9}$ centimeters per second. For sanitary landfills material of this kind is ideal if it has relatively few areas which are interbedded with sand and gravel and is readily accessible.

Fortunately a publication of the Illinois State Geological Survey (ISGS) contains this information. ISGS Circular Number 532 entitled "Potential for Contamination of Shallow Aquifers in Illinois" contains a map which identifies uniform, relatively impermeable silty or clayey till at least 50 feet thick with no evidence of interbedded sand or gravel.

## Surface Waters

Surface waters represent a valuable natural resource which might be at risk from water pollution if the surface flow across a landfill site is not properly handled. The design and the operating plan seek to control this flow, but as an added precaution a reasonable separation from surface water resources is a prudent measure.

A current proposal by the Illinois Pollution Control Board prohibits landfill siting within 600 feet of a navigable lake or pond and within 300 feet of a river or stream. The more conservative 600 feet separation was used for all surface water resources. Data for this map were provided by the Lake County Department of Management Services.

## Surficial Aquifers

Like surface waters, surficial aquifers (water bearing strata near the ground surface) are particularly vulnerable to pollution from improperly managed drainage from sanitary landfills. Surficial aquifers were therefore identified as areas to be excluded from consideration. Since runoff control is also a concern in a regional composting operation the surficial aquifers were also identified as exclusion zones for the composting site selection map.

The data for mapping surficial aquifers were identified using ISGS Circular 481 - Geology for Planning in Lake County.


## CULTURAL/INSTITUTIONAL RESOURCES

## Historical Resources

The Illinois Pollution Control Board currently proposes that solid waste disposal facilities be restricted from development in locations which might pose a threat to irreplaceable historical resources.

The Illinois Historic Structures Survey was used as the data base and the locations were plotted with a 200 feet exclusion zone. This exclusion applied to all solid waste management facilities.

## Archaeological/Paleontological

The above proposed regulation also covers archaeological and paleontological resources. However, the Illinois Archaeological Survey declined to make this information available on the grounds that it could only be interpreted by an archaeologist. During subsequent phases of the siting process, archaeological and paleontological information should be requested for specific areas of the county judged to be suitable for regional pollution control facilities.

Schools/Hospitals

The Illinois Pollution Control Board (IPCB) currently proposes a 500 foot exclusion zone around schools and hospitals. These institutions were plotted and identified as exclusion areas for all solid waste management facilities.

## DEVELOPMENT

### Existing Development

Sanitary landfills and regional composting facilities were judged to be necessarily excluded from proximity to existing development. Since the IPCB proposes 500 foot separation this figure was used to arrive at the exclusion zone.

For waste-to-energy facilities a similar exclusion exists for residential and commercial development. However, a waste-to-energy facility might reasonably be sited in an industrial area; in fact, under some circumstances adjacent industries might be potential energy users. Since industrial land uses were not available in a convenient data base it was decided not to add existing development as a specific exclusion for the waste-to-energy map. Clearly, this leaves a larger potential site area for waste-to-energy facilities than is strictly consistent with the criteria. As a first level approximation this is justified on the grounds that a later evaluation will exclude residential and commercial areas within 500 feet of a potential site.

Transfer station sites should also be excluded from within 500 feet of residential and commercial property but could be included in industrial areas. For the same reason discussed above, existing development was not mapped as an exclusion area for transfer stations.

In addition, it was necessary to identify a maximum lot size above which residential areas would cease to be considered development. For example, a single house on a 2 acre property represents 2 acres of existing development, all of which is therefore excluded. However, a single house on 10 acres, especially if the house is in the corner of the property, does not necessarily represent 10 acres of existing development. It was decided to use 5 acres as the dividing line.

### Committed Development

Any real estate which has been committed to development should also be excluded since the committed development is almost certain to be consummated by the time a regional solid waste management facility is sited.

For reasons discussed above existing and committed development is mapped as an exclusion for sanitary landfill and composting and not for waste-to-energy and transfer stations.

There are many different notions of what committed development really means so it was necessary to adopt a working definition. The siting committee decided that committed development would be defined as development which has received final plat approval and has been active within the last two years.

## PROXIMITY TO MAJOR HIGHWAYS

### Proximity to Major Highways

All the solid waste management systems currently available have one thing in common and that is dependance upon road access. Since most solid waste is transported by packer truck it is important that well constructed roads should be used with at least an 8 ton per axle design loading. Thus, one criterion is proximity to county roads with at least 8 ton per axle design loading. State roads which are generally constructed to a 9 ton axle load design should also be included.

Given that the packer truck has access to roads of the appropriate design loading, it is important to determine how much new road must be constructed from the highway to the facility itself. This access road must be constructed to equally high standards. The road will be costly and should therefore be as short as possible. However, land adjacent to major highways is more expensive than land for which an additional access road must be constructed.

Since we propose to map all areas of suitable location the question is - what is the longest acceptable access road to each of the potential elements of the solid waste management system.

For sanitary landfill the access road itself should not be long, although at least 250-500 feet should be available to allow the vehicle to remove mud collected on the tires after tracking over the landfill surface. However, the landfill itself takes up a significant land area and an additional road is necessary across the center of the landfill to provide access to each cell. The largest landfill necessary to meet all of Lake County's needs for the design period of 20 years would need to be capable of accepting an average of 2000 tons of solid waste per day. Conceptually, this could be accomplished in a landfill measuring 3200 x 3200 feet. The road across the center of this area would be 3200 feet in length. If the original maximum of 500 feet of access road is added to 3200 feet, a maximum distance of 3700 feet from the major highway is calculated.

For waste-to-energy facilities, approximately half a mile of access road is acceptable. The exclusion area was therefore identified as the area beyond 2500 feet of major highways.

For transfer stations an access road not to exceed 500 feet is acceptable. Composting operations were not considered to be constrained by access road distance since the axle loads associated with vehicles carrying yard waste are not as high as those carrying solid waste, in addition to which the traffic level is much lower than that encountered for the other types of operation.

# ADDITIONAL FEATURES

## Airports

The Federal Aviation Authority restricts the construction of a sanitary landfill within 5000 feet of an airport used by turboprop aircraft and within 10,000 feet of an airport used by jet aircraft. The two airports in Lake County were mapped and the 5000 feet radius circles plotted as exclusion zones.

## Municipal Wells

The newly enacted State Groundwater Protection Act prohibits construction of a sanitary landfill within 500 feet of a municipal water supply.

The members of the Joint Action Agency were requested to provide the locations of all municipal supply wells. These were plotted and the 500 feet radius areas identified as an exclusion zones.

## THE SIEVING PROCESS

Each of the above criteria were mapped by the Illinois Department of Energy and Natural Resources using advanced digitizing equipment and the extensive computer mapping capability of the Department.

In the days before computer mapping each map was drawn on clear acetate and placed on top of the previous one until finally from the complete set of maps the most suitable areas could be "sieved out". Thanks to the computer a composite map can be produced very easily once all the data have been incorporated.

The following table identifies which data were used to produce the composite map for the sanitary landfill option.

## TABLE 9-1
## LANDFILL SITING CRITERIA

| Criteria | Sanitary Landfill |
|---|---|
| **Natural Features** | |
| 1. Flood Plain | X |
| 2. Wetlands | X |
| 3. Threatened & Endangered Species | X |
| 4. Suitable soils | X |
| 5. Surface waters (600 ft) | X |
| 6. Surficial aquifers | X |
| **Cultural/Institutional Features** | |
| 1. Historic/ Architectural (200 ft) | X |
| 2. Archaeological (200 ft) | X |
| 3. Paleontological (200 ft) | X |
| 4. Schools/Hospitals (500 ft) | X |
| **Development** | |
| 1. Existing (500 ft) | X |
| 2. Committed (500 ft) | X |
| **Proximity to Major Highways** | X [1] |
| **Additional Features** | |
| 1. Airports (5000 ft) | X |
| 2. Municipal wells (500 ft) | X |

1. 3700 feet

138

## COMPOSITE MAPS

### Sanitary Landfill Composite

The maps used to generate the composite map for landfill siting are presented as follows:

> Figure 9-1 Natural Resources.  Seive 1
> Figure 9-2 Cultural/Institutional Resources.  Seive 2
> Figure 9-3 Development.  Seive 3
> Figure 9-4 Proximity to Major Highways.  Seive 4
> Figure 9-5 Landfill Composite Map.

The sanitary landfill composite map is presented as Figure 9.5.  It should be noted that all areas less than 60 acres have been excluded since it is estimated that the smallest sanitary landfill to serve the area would be 60 acres.

A separate map was produced for the additional features, but it is not reproduced here as it adds little to an understanding of the overall process.  However, these data were used in the development of the composite map.

The purpose of the composite maps is to identify the development of the final map in manageable sets of data.

By progressing through each seive map, the reader can identify the development of the final map of suitable areas.

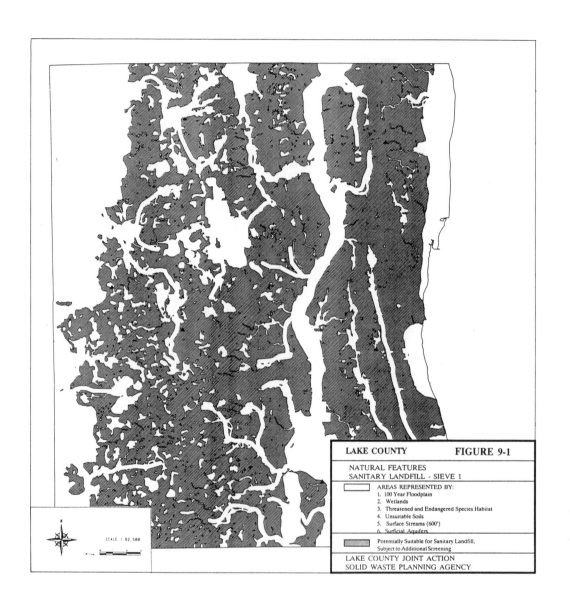

LAKE COUNTY                    FIGURE 9-1

NATURAL FEATURES
SANITARY LANDFILL - SIEVE 1

AREAS REPRESENTED BY:
1. 100 Year Floodplain
2. Wetlands
3. Threatened and Endangered Species Habitat
4. Unsuitable Soils
5. Surface Streams (600')
6. Surficial Aquifers

Potentially Suitable for Sanitary Landfill,
Subject to Additional Screening

LAKE COUNTY JOINT ACTION
SOLID WASTE PLANNING AGENCY

SCALE 1:62,500

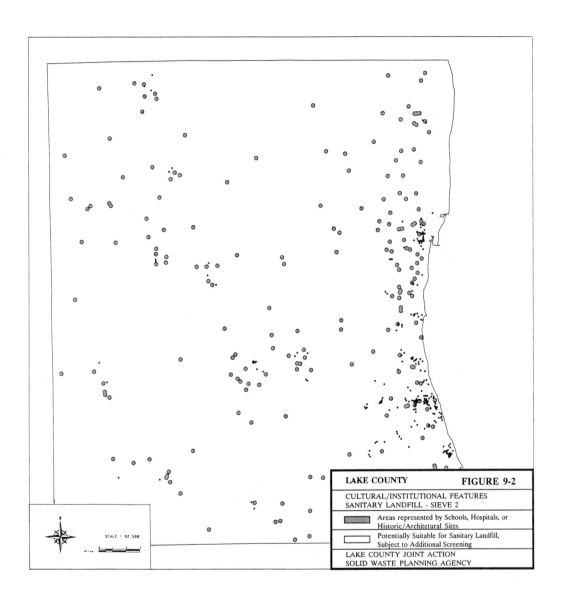

LAKE COUNTY                              FIGURE 9-2

CULTURAL/INSTITUTIONAL FEATURES
SANITARY LANDFILL - SIEVE 2

Areas represented by Schools, Hospitals, or
Historic/Architetural Sites

Potentially Suitable for Sanitary Landfill,
Subject to Additional Screening

LAKE COUNTY JOINT ACTION
SOLID WASTE PLANNING AGENCY

SCALE 1 62,500

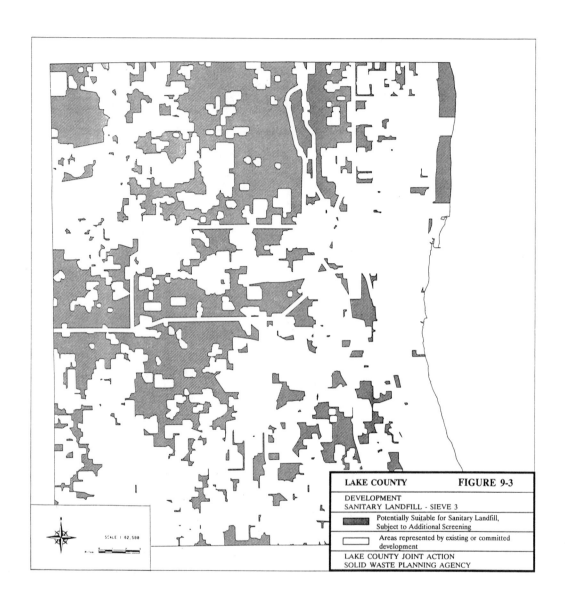

LAKE COUNTY                    FIGURE 9-3

DEVELOPMENT
SANITARY LANDFILL - SIEVE 3

Potentially Suitable for Sanitary Landfill,
Subject to Additional Screening

Areas represented by existing or committed
development

LAKE COUNTY JOINT ACTION
SOLID WASTE PLANNING AGENCY

SCALE 1 62,500

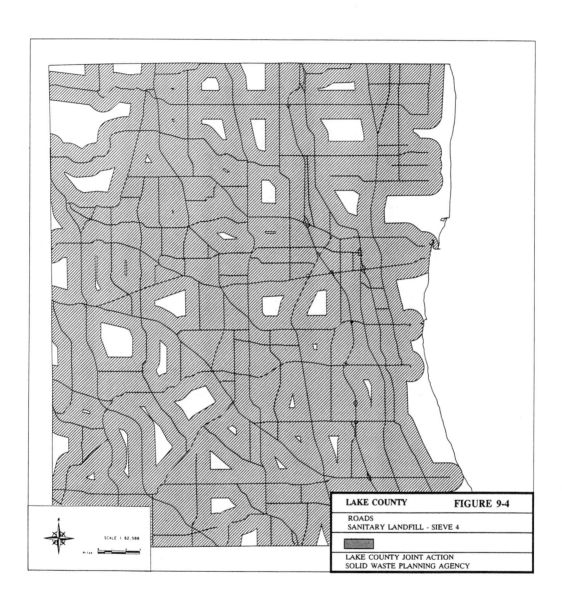

LAKE COUNTY                    FIGURE 9-4

ROADS
SANITARY LANDFILL - SIEVE 4

LAKE COUNTY JOINT ACTION
SOLID WASTE PLANNING AGENCY

SCALE 1 62,500

143

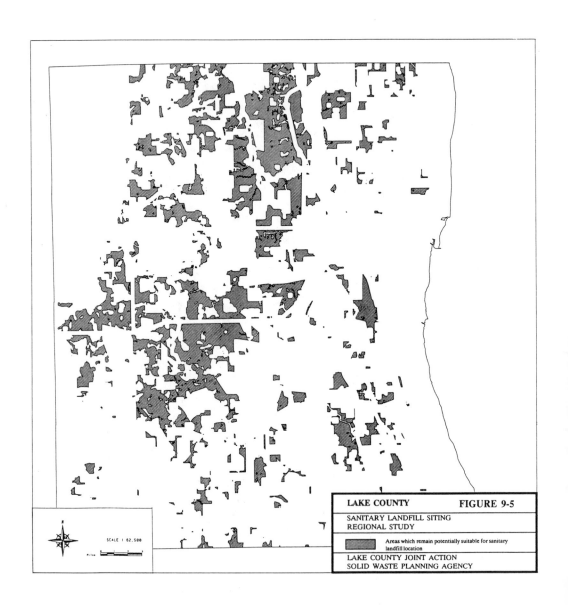

LAKE COUNTY          **FIGURE 9-5**

SANITARY LANDFILL SITING
REGIONAL STUDY

Areas which remain potentially suitable for sanitary landfill location

LAKE COUNTY JOINT ACTION
SOLID WASTE PLANNING AGENCY

SCALE 1 62,500

144

# 10

# Siting Criteria for Other LULUs

## INTRODUCTION

There are many examples of Locally Unacceptable Land Uses or LULUs to be found. Those discussed in this book are devoted to pollution control, although there are many examples from other industries which could be sited in the same manner.

Those discussed in this book are:

> Landfills
> Transfer Stations
> Incinerators
> Compost Facilities
> Sewage Treatment Facilities
> Recycling Facilities

It is a central thesis of this book that the siting procedure for each of these types of development should follow that described in detail for landfills. The proper application of appropriate siting criteria in the two step regional and local process provides the best possible opportunity to achieve a successful siting.

The process of developing siting criteria for transfer stations can be handled as in the landfill case.

## TRANSFER STATIONS

The first step is to develop a matrix of environmental impacts based on the various operational phases of transfer station operation. First review transfer station operation, and identify the various operational phases. It should be noted that unlike landfill operation in which the actual construction phase is an integral part of the operation, the transfer station has a well defined construction phase. In this case, and in most other LULUs, the matrix should be expanded to cover the construction phase.

## TABLE 10-1
### CONSTRUCTION AND OPERATIONAL PHASES
### OF TRANSFER STATION DEVELOPMENT

Excavation
Fill
Building
Haul to Transfer
Queue at Station
Garbage Deposition
Push to Compactor
Compaction Cycle
Discharge to Trailer
Haul to Disposal

A suggested list of construction and operational phases necessary in the development of a transfer station is presented in Table 10-1.

Next review these operational phases, and identify the attendant environmental impacts. If the impact factors suggest new subcategories of construction or operation, go back and make the change in the first step.

## TABLE 10-2
### ENVIRONMENTAL IMPACTS ASSOCIATED
### WITH TRANSFER STATION DEVELOPMENT

Use of Land
Highway Congestion
Highway Noise
Accidents
Noise
Air Pollution
Rats
Flies
Vibration
Groundwater Contamination
Surface Water Contamination
Economic Loss

Now mount these two lists together on a grid to form a matrix as in the landfill example in Chapter 3, and as shown in Table 10-3. Review each phase of the construction and operation, and mark those items which will give rise to specific environmental impacts.

**TABLE 10-3**
**MATRIX OF CONTRUCTION AND OPERATIONAL PHASES AND ENVIRONMENTAL IMPACTS FOR TRANSFER STATIONS**

Environmental Impacts

| Construction and Operational Phases | Use of land | Highway congestion | Highway noise | Accidents | Noise | Air Pollution | Rats | Flies | Vibration | Groundwater contamination | Surface water contamination | Economic loss |
|---|---|---|---|---|---|---|---|---|---|---|---|---|
| Excavation | X | | | | X | X | | | | | | |
| Fill | X | | | | X | X | | | | | | |
| Building | X | | | | X | | | | | | | |
| Haul to transfer | | X | X | X | | | | | | | | |
| Queue at station | | X | | | | | | | | | | |
| Garbage deposition | | | | | X | X | | | | | | |
| Push to compactor | | | | | X | | | | | | | |
| Compaction cycle | | | | | X | | | | | | X | |
| Discharge to trailer | | | | | X | | X | | | | | |
| Haul to disposal | | X | X | X | | | | | | | | |

147

Clearly many of the environmental risks depend upon specific circumstances. For example, whether or not there is highway congestion during the filling phase depends upon how much fill is imported to the site. Since transfer stations by their design require two levels, the higher level for incoming loads, and the lower level for outgoing compacted loads, there are essentially two kinds of design. One design will involve excavation using the excavated material to raise the ramp, and the operating floor to the required level. A second design will import fill material from off site to achieve this raised section without excavation. Clearly the individual elements of the construction phase depend upon which design is selected.

For many years, it has been obvious that if an enterprising designer could develop a single story transfer operation, the entire siting, permitting and construction phases would be greatly simplified. To date this has not happened, and the world of waste management still waits for the single level transfer station.

With the first matrix completed, we can now proceed to identify the specific environmental impacts associated with transfer station development as shown in Table 10-4.

Basically, we have used Table 10-3 as a tool to identify potential environmental impacts. Now we need to identify the various sensitive elements of the environment which could be impacted by these risks. Starting with the same list that was used in developing landfill siting criteria, we can now build a list of sensitive environments appropriate to transfer station development as shown in Table 10-5.

TABLE 10-4
POTENTIAL ENVIRONMENTAL IMPACTS
OF TRANSFER STATION DEVELOPMENT

Use of Land for Excavation
Use of Land for Filling
Use of Land for Building
Highway Congestion Haul to Transfer
Highway Congestion Queue at Station
Highway Congestion Haul to Disposal
Highway Noise Due to Haul
Accidents Due to Haul
Noise from Haul
Noise for Garbage Deposition
Noise From Compaction
Air Pollution From Building
Air Pollution From Garbage Deposition
Vibration from Compaction
Air Pollution From Queuing at Station

TABLE 10-5
ENVIRONMENTS SENSITIVE
TO TRANSFER STATION DEVELOPMENT

Sensitive Environments

Wetlands
Flood Plains
Surface Waters
Groundwater Resources
Threatened & Endangered Species Habitat
Scenic Areas
Existing Depressions
Natural Screening

Land Use

Development (Existing or Committed)
Areas of Historic Importance
Areas of Architectural Importance
Areas of Paleontological Importance
Areas of Archaeological Importance
Areas with Natural Buffering
Proximity to Municipal Boundaries

Economic Factors

Proximity to Major Highways
Highway Restrictions
Traffic
Distance From Centroid of Waste Generation or Transfer
Availability

Table 10-6 shows how this second matrix is constructed with specific environmental impacts across the top, and specific sensitive elements of the environment down the vertical axis. The list of sensitive elements is essentially generic, and should be expanded in areas where other features of the natural environment may be at risk.

By taking each impact in turn, and travelling down the matrix, that element of the environment which is most affected by the impact can be identified.

**TABLE 10-6**
MATRIX OF ENVIRONMENTAL IMPACTS OF TRANSFER STATION
DEVELOPMENT AND POTENTIAL SITING CRITERIA

**Environmental Impacts**

| Sensitive environments | Use of land for excavation | Use of land for filling | Use of land for building | Highway congestion, haul to transfer | Highway congestion, queue at station | Highway congestion, haul to disposal | Highway noise due to haul | Accidents due to haul | Noise from haul | Noise from garbage deposition | Noise from compaction | Air pollution from building | Air pollution from garbage deposition | Vibration from compaction | Air pollution from queueing at station |
|---|---|---|---|---|---|---|---|---|---|---|---|---|---|---|---|
| Wetlands | | X | | | | | | | | | | | | | |
| Flood Plains | | | X | | | | | | | | | | | | |
| Surface Waters | | | | | | | | | | | | | | | |
| Groundwater Resources | | | | | | | | | | | | | | | |
| Threatened & endangered spec. habitat | X | X | X | | | X | | | | X | X | X | X | X | X |
| Scenic Areas | | | X | | | | | | | | | | | | |
| Existing depressions | | | | | | | | | | | | | | | |
| Natural screening | | | X | | | | | | | | | | | | |

150

TABLE 10-6, Continued

**Environmental Impacts**

| Land Use | Use of land for excavation | Use of land for filling | Use of land for building | Highway congestion, haul to transfer | Highway congestion, queue at station | Highway congestion, haul to disposal | Highway noise due to haul | Accidents due to haul | Noise from haul | Noise from garbage deposition | Noise from compaction | Air pollution from building | Air pollution from compaction | Air pollution from garbage deposition | Vibration from compaction | Air pollution from queueing at station |
|---|---|---|---|---|---|---|---|---|---|---|---|---|---|---|---|---|
| Development (Existing or Committed | | | | X | X | X | X | X | X | X | X | X | X | X | X | X |
| Airports (Jet or Turboprop) | | | | | | | | | | | | | | | | |
| Prime Farmland | X | X | X | | | | | | | | | | | | | |
| Municipal water supply wells | | | | | | | | | | | | | | | | |
| Areas of historic importance | | | X | | | X | | | | | X | X | X | X | X | X |
| Areas of architectural importance | | | X | | | X | | | | | X | X | X | X | X | X |
| Areas of paleontological importance | X | | | | | | | | | | | | | | | |
| Areas of archaeological importance | X | | | | | | | | | | | | | | | |
| Areas with natural buffering | | | X | | | | | | | | | | | | | |
| Proximity to areas with final use compatibility | | | | | | | | | | | | | | | | |
| Proximity to municipal boundaries | | | X | | | | | | | | | | | | | |

151

TABLE 10-6, Continued

## Environmental Impacts

| Economic Factors | Use of land for excavation | Use of land for filling | Use of land for building | Highway congestion, haul to transfer | Highway congestion, queue at station | Highway congestion, haul to disposal | Highway noise due to haul | Accidents due to haul | Noise from haul | Noise from garbage deposition | Noise from compaction | Air pollution from building | Air pollution from garbage deposition | Vibration from compaction | Air pollution from queueing at station |
|---|---|---|---|---|---|---|---|---|---|---|---|---|---|---|---|
| Proximity to major highways | | | | X | | | | | | | | | | | |
| Highway restrictions | | | | X | | | | | | | | | | | |
| Traffic | | | | X | | | | | | | | | | | |
| Distance from centroid of waste generation or transfer | | | | X | | | | | | | | | | | |
| Availability | X | X | X | | | | | | | | | | | | |
| Land holding in large parcels | X | X | X | | | | | | | | | | | | |

Using Table 10-6 as shown, the siting criteria for transfer station development can now be identified as follows:

TABLE 10-7
SITING CRITERIA FOR TRANSFER STATION DEVELOPMENT

Wetlands Exclusion
Flood Plain Exclusion
Threatened and Endangered Species Habitat Exclusion
Scenic Area (Sxclusion or Screening)
Natural Screening Inclusion
Development Exclusion (Residential Only)
Prime Farmland Exclusion
Historic Area Exclusion
Architectural area exclusion
Paleontological Area Exclusion
Archaeological Area Exclusion
Proximity to Highway Inclusion
Highway Restriction Exclusion
Congested Area Exclusion
Minimize Distance From Centroid of Waste Generation
Available Land Inclusion
Land Holding in Large Parcel Inclusion

Review of these siting criteria will generate a regional and a local list of siting criteria. Remember that the application of the regional criteria is intended to reduce the area of the search, and the local criteria are required to identify a list of sites.

The regional siting criteria suggested are presented in Table 10-8.

TABLE 10-8
REGIONAL SITING CRITERIA
FOR TRANSFER STATION DEVELOPMENT

Wetlands
Flood Plains *
Proximity to Major Highway **

\*      A minimum distance of 500 feet is suggested

\*\*     A maximum distance of 500 feet is suggested

153

TABLE 10-9
LOCAL SITING CRITERIA
FOR TRANSFER STATION DEVELOPMENT

Threatened and Endangered Species Habitat Exclusion
Scenic Area (Exclusion or Screening)
Natural Screening Inclusion
Development Exclusion (Residential Only)
Prime Farmland Exclusion
Historic Area Exclusion
Architectural Area Exclusion
Paleontological Area Exclusion
Archaeological Area Exclusion
Highway Restriction Exclusion
Congested Area Exclusion
Minimize Distance From Centroid of Waste Generation
Available Land Inclusion
Land Holding in Large Parcel Inclusion

The local siting criteria suggested are presented in Table 10-9.

INCINERATORS

The construction and operational phases of incinerator development are identified in Table 10-10.

TABLE 10-10
CONSTRUCTION AND OPERATIONAL PHASES
OF INCINERATOR DEVELOPMENT

Excavation
Fill
Building
Haul to Incinerator
Queue at Incinerator
Garbage Disposition
Push to Charging Pit
Combustion
Gaseous Emission
Steam Generation
Ash Production
Ash Haul
Ash Disposal

The environmental impacts associated with these phases of construction and operation are as follows:

## TABLE 10-11
## ENVIRONMENTAL IMPACTS ASSOCIATED
## WITH INCINERATION DEVELOPMENT

Highway Congestion
Highway Noise
Accidents
Noise
Air Pollution
Economic Loss
Reduction of Scenic Quality

Adding these two lists together as in previous examples produces a matrix of construction and operational phases of incineration development versus environmental impacts

# TABLE 10-12
## MATRIX OF CONSTRUCTIONAL AND OPERATIONAL PHASES OF INCINERATION DEVELOPMENT VERSUS ENVIRONMENTAL IMPACTS

**Environmental Impacts**

| Construction and Operational Phases | Highway congestion | Highway noise | Accident | Noise | Air Pollution | Economic Loss | Reduction of Scenic Quality |
|---|---|---|---|---|---|---|---|
| Excavation | | | | X | | | |
| Fill | | | | X | | | |
| Building | | | | X | | | X |
| Haul to incinerator | X | X | X | | | | |
| Queue at incinerator | X | | | | | X | |
| Garbage deposition | | | | X | | | |
| Push to charging pit | | | | X | | | |
| Combustion | | | | | | | |
| Gaseous emission | | | | | X | | |
| Steam generation | | | | | | | X |
| Ash production | | | | | X | | |
| Ash haul | X | X | X | | X | | |
| Ash disposal | | | | | X | | |

156

TABLE 10-13
POTENTIAL ENVIRONMENTAL IMPACTS
OF INCINERATOR DEVELOPMENT

Use of Land for Building
Highway Congestion/Haul to Incinerator
Highway Congestion/Queue
Highway Congestion/Ash Haul
Highway Noise/Haul to Incinerator
Highway Noise/Ash Haul
Accident/Haul to Incinerator
Accident/Ash Haul
Noise/Excavation
Noise/Fill
Noise/Building
Noise/Garbage Deposition
Noise/Push to Charging Pit
Air Pollution/Gas Emission
Air Pollution/Ash Production
Air Pollution/Ash Haul
Air Pollution/Ash Disposal
Economic Loss/Queue at Incinerator
Reduction of Scenic Quality/Building
Reduction of Scenic Quality/Steam Generation
Overall Economic Loss or Gain

With this first matrix completed, we can identify potential environmental impacts which may occur as a result of one or more aspects of incinerator development.

Which environments provide the best protection against these impacts? Which environments are most sensitive to these types of impacts? These are the questions to be answered in the development of the second matrix.

As before the next step is to identify the potentially sensitive environments starting with the list developed earlier for landfills, and tailoring this to meet incineration related environments.

## TABLE 10-14
## POTENTIALLY SENSITIVE ENVIRONMENTS
## AND OTHER FACTORS LIMITING INCINERATION DEVELOPMENT

Wetlands
Flood Plains
Threatened & Endangered Habitats
Scenic Areas
Natural Screening
Development (Exist & Committed)
Areas of Historic Importance
Areas of Architectural Importance
Areas of Paleontological Importance
Areas of Archaeological Importance
Natural Buffering
Close to Major Highway
Highway Restrictions
Close to Centroid
Available Land

The second matrix can now be developed as shown in Table 10-15.

**Environmental Impacts of Incineration Development**

| Sensitive Environments | Use of land for building | Highway congestion / Haul | Highway congestion / Queue | Highway congestion / Ash haul | Highway noise / Haul | Highway noise / Ash haul | Accidents / Haul | Accidents / Ash haul | Noise / Excavation | Noise / Fill | Noise / Building | Noise / Garbage deposit | Noise / Push | Air pollution / Gas | Air pollution / Ash | Air pollution / Ash haul | Economic loss / Queue | Reduction of scenic quality / Ash disposal | Reduction of scenic quality / Building | Overall Economic Loss or Gain / Steam gen |
|---|---|---|---|---|---|---|---|---|---|---|---|---|---|---|---|---|---|---|---|---|
| Wetlands | X | | | | | | | | | | | | | | | | | | | |
| Flood Plains | X | | | | | | | | | | | | | | | | | | | |
| Threatened and endangered species habitat | X | | | | | | | | | X | X | | | X | X | X | X | | | |
| Scenic Areas | | | | | | | | | | | | | | | | | | | | |
| Natural Screening | | | | | | | | | | | | | | | | | | X | | |
| Development (Existing & Comm.) | | X | | | | | X | X | X | X | X | X | X | X | X | X | X | X | X | |
| Areas of historic importance | | X | | | | | | | | | | | | | | | | | X | X |
| Areas of architectural importance | | X | | | | | | | | | | | | | | | | | X | X |
| Areas of paleontological importance | | X | | | | | | | | | | | | | | | | | | |
| Areas of archaeological importance | | X | | | | | | | | | | | | | | | | | | |
| Natural Buffering | | | | | | | | | | X | X | X | X | | | | | | | |
| Proximity to major highway | | | | | | | | | | | | | | | | | | | | X |
| Highway Restrictions | | X | | | | | | | | | | | | | | | | | | X |
| Proximity to centroid | | | | | | | | | | | | | | | | | | | | X |
| Availability of land | | | | | | | | | | | | | | | | | | | | X |

159

## TABLE 10-16
## SELECTION CRITERIA FOR INCINERATION DEVELOPMENT

Wetlands
Flood Plains
Threatened & Endangered Habitats
Scenic Areas
Natural Screening
Development (Exist & Committed)
Areas of Historic Importance
Areas of Architectural Importance
Areas of Paleontological Importance
Areas of Archaeological Importance
Natural Buffering
Close to Major Highway
Highway Restrictions
Close to Centroid
Available Land

From the matrix shown in Table 10-15, we can develop the final list of selection criteria as shown in Table 10-16.

Now we need to divide this list into regional and local criteria.

The local siting criteria are presented in Table 10-18.

## TABLE 10-17
## REGIONAL SITING CRITERIA
## FOR INCINERATION DEVELOPMENT

Wetlands
Flood Plains *
Proximity to Major Highways **

\* A minimum distance of 500 feet is suggested.

\** A maximum distance of 500 feet is suggested.

## TABLE 10-18
## LOCAL SITING CRITERIA
## FOR INCINERATION DEVELOPMENT

Threatened & Endangered Habitats
Scenic Areas
Natural Screening
Development (Exist & Committed)
Areas of Historic Importance
Areas of Architectural Importance
Areas of Paleontological Importance
Areas of Archaeological Importance
Natural Buffering
Close to Major Highway
Highway Restrictions
Close to Centroid
Available Land

## COMPOST FACILITIES

The construction and operational phases of compost facility development are presented in Table 10-19.

Next review these events, and develop a list of attendant environmental impacts. If the impacts suggest new subcategories of construction or operation, go back, and add them to the first list.

## TABLE 10-19
## CONSTRUCTION AND OPERATIONAL PHASES
## OF COMPOST FACILITY DEVELOPMENT

Preparation of Compost Pad
Haul to Facility
Queue at Facility
Deposition
Turnover
Watering
Removal of Composted Material
Haul to Point of Use

TABLE 10-20
ENVIRONMENTAL IMPACTS ASSOCIATED
WITH COMPOST FACILITY DEVELOPMENT

Use of Land
Highway Congestion
Highway Noise
Accidents
Noise
Air Pollution
Flies
Groundwater Contamination
Economic Loss

Now we can construct the first matrix in order to identify the potential risks of compost facility development

TABLE 10-21:

# MATRIX OF CONSTRUCTION AND OPERATIONAL PHASES AND THE ENVIRONMENTAL IMPACTS OF COMPOST FACILITIES

**Environmental Impacts**

| Construction and Operational Phases | Use of land | Highway congestion | Highway noise | Accidents | Noise | Air Pollution | Flies | Groundwater contamination | Economic loss |
|---|---|---|---|---|---|---|---|---|---|
| Preparation of compost pad | X | | | | | | | | |
| Haul to facility | | X | X | X | | | | | |
| Queue at facility | | | | | X | | | | X |
| Deposition | | | | | | | | | |
| Turnover | | | | | | X | | | |
| Watering | | | | | | | | X | |
| Removal | | | | | | X | | | |
| Haul | | | | | | | | | |

163

With the first matrix completed, we can now identify specific impacts associated with specific aspects of construction and operation as shown in Table 10-22.

TABLE 10-22
POTENTIAL ENVIRONMENTAL IMPACTS
OF COMPOST FACILITY DEVELOPMENT

Use of Land for Compost Pad
Highway Congestion/Haul to Facility
Highway Noise/Haul to Facility
Accident Potential/Haul to Facility
Noise/Queue at Facility
Air Pollution/Turnover of Compost
Air Pollution/Removal of Compost
Groundwater Contamination/Watering
Economic Loss/Queue at Facility

Now we need to identify the various sensitive environmentals which could be impacted by these potential risks. Starting with the same list that was used in developing landfill siting criteria, we can build a list appropriate to composting as shown in Table 10-23.

TABLE 10-23
ENVIRONMENTS SENSITIVE TO
COMPOST FACILITY DEVELOPMENT

Sensitive Environments

    Wetlands
    Flood Plains
    Surface Waters
    Groundwater Resources
    Threatened & Endangered Species Habitat
    Scenic Areas
    Natural Screening

Land Use

    Development (Existing or Committed) Except Industrial
    Areas of Historic Importance
    Areas of Architectural Importance
    Areas of Paleontological Importance
    Areas of Archaeological Importance
    Proximity to Municipal Boundaries

Economic Factors

    Proximity to Major Highways
    Highway Restrictions
    Traffic
    Distance from Centroid of Waste Generation
    Availability

Now we can construct the second matrix which will be used to identify siting criteria.

TABLE 10-24:

MATRIX OF ENVIRONMENTAL IMPACTS OF
COMPOST FACILITY DEVELOPMENT
AND POTENTIAL SITING CRITERIA

**Environmental Impacts**

| | Use of land for compost pad | Highway congestion / Haul to facility | Highway noise / Haul to facility | Accident potential / Haul to facility | Noise / Queue at facility | Air pollution / Turnover of compost | Air pollution / Removal of compost | Groundwater contamination / Watering | Economic loss / Queue at facility |
|---|---|---|---|---|---|---|---|---|---|
| **Sensitive Environments** | | | | | | | | | |
| Wetlands | X | | | | | | | | |
| Flood Plains | X | | | | | | | | |
| Surface Waters | | | | | | | | | |
| Groundwater Resources | | | | | | | | X | |
| Threatened & endangered species habitat | X | | | | | | | | |
| Scenic Areas | X | | | | | | | | |
| Natural Screening | | | | | X | | | | |
| **Land Use** | | | | | | | | | |
| Development (existing or committed), except industrial | | | | | X | X | X | | X |
| Areas of historic importance | | X | | | | | | | |
| Areas of architectural importance | | X | | | | | | | |
| Areas of paleontological importance | | X | | | | | | | |
| Areas of archaeological importance | | X | | | | | | | |

166

TABLE 10-24, Continued

**Environmental Impacts**

| Economic Factors | Use of land for compost pad | Highway congestion / Haul to facility | Highway noise / Haul to facility | Accident potential / Haul to facility | Noise / Queue at facility | Air pollution / Turnover of compost | Air pollution/ Removal of compost | Groundwater contamination / Watering | Economic loss/ Queue at facility |
|---|---|---|---|---|---|---|---|---|---|
| Proximity to major highways | | | | | | | | | X* |
| Highway Restrictions | | X | | | | | | | |
| Traffic | | X | | | | | | | |
| Distance from centroid of waste generation | | X | X | X | | | | | |
| Availability | X | | | | | | | | |

* Actually there is an economic gain due to a reduction in the length of haul road required.

167

Based on the results of this matrix, we can now identify the siting criteria for siting compost facilities.

TABLE 10-25
SITING CRITERIA
FOR COMPOST FACILITIES

Sensitive Environments

Wetlands
Flood Plains
Surface Waters
Groundwater Resources
Threatened & Endangered Species Habitat
Scenic Areas
Natural Screening

Land Use

Development (Existing or Committed) Except Industrial
Areas of Historic Importance
Areas of Architectural Importance
Areas of Paleontological Importance
Areas of Archaeological Importance
Proximity to Municipal Boundaries

Economic Factors

Proximity to Major Highways
Highway Restrictions
Traffic
Distance from Centroid of Waste Generation
Availability

Now we can proceed to identify the regional criteria.

## TABLE 10-26
## REGIONAL CRITERIA FOR COMPOST FACILITY SITING

Wetlands
Flood Plains
Surface Waters*
Groundwater Resources
Development (Existing & Committed) Except Industrial
Proximity to Major Highways**

\*  500 feet exclusion distance is suggested
\*\* 500 feet maximum distance is suggested

The local siting criteria list is all that remains.

## TABLE 10-27
## LOCAL CRITERIA FOR COMPOST FACILITY SITING

Threatened & Endangered Species Habitat
Scenic Area
Areas of Historic Importance
Areas of Architectural Importance
Areas of Paleontological Importance
Areas of Archaeological Importance
Proximity to Municipal Boundaries
Highway Restrictions
Traffic
Distance From Centroid of Waste Generation
Availability

TABLE 10-28
CONSTRUCTION AND OPERATIONAL PHASES
OF SEWAGE TREATMENT FACILITY

Excavation
Fill
Building
Treatment Operations
Air Emissions
Sludge Production
Sludge Haul

The environmental impacts associated with these phases of construction and operation are as follows:

TABLE 10-29
ENVIRONMENTAL IMPACTS ASSOCIATED
WITH SEWAGE TREATMENT FACILITY DEVELOPMENT

Use of Land
Highway Congestion
Highway Noise
Accidents
Noise
Air Pollution
Groundwater Contamination
Economic Loss
Reduction of Scenic Quality

Adding these two lists together as in previous examples produces a matrix of construction and operational phases of sewage treatment facility development versus environmental impacts.

TABLE 10-30:

## MATRIX OF CONSTRUCTION AND OPERATIONAL PHASES OF SEWAGE TREATMENT FACILITIES VERSUS ENVIRONMENTAL IMPACTS

**Environmental Impacts**

| Construction and Operational Phases | Use of land | Highway congestion | Highway noise | Accidents | Noise | Air Pollution | Groundwater contamination | Economic loss | Reduction of Scenic Quality |
|---|---|---|---|---|---|---|---|---|---|
| Excavation | | | | | X | | | | |
| Fill | | | | | X | | | | |
| Building | X | | | | X | | | | X |
| Treatment Operations | | | | | | X | X | X | |
| Air Emissions | | | | | | X | | | |
| Sludge Production | | | | | | | | | |
| Sludge Haul | | X | X | X | X | X | | | |

TABLE 10-31
POTENTIAL ENVIRONMENTAL IMPACTS
OF SEWAGE TREATMENT FACILITY DEVELOPMENT


Use of Land for Building
Highway Congestion/Sludge Haul
Highway Noise/Sludge Haul
Accident/Sludge Haul
Noise/Excavation
Noise/Fill
Noise/Building
Noise/Sludge Haul
Air Pollution/Treatment Operation
Air Pollution/Air Emissions
Air Pollution/Sludge Haul
Economic Loss/Treatment Operations
Reduction of Scenic Quality/Building


With this first matrix completed, we can identify potential environmental impacts which may occur as a result of one or more aspects of sewage treatment facility development.

Which environments provide the best protection against these impacts? Which environments are most sensitive to these types of impacts? These are the questions to be answered in the development of the second matrix.

As before the next step is to identify the potentially sensitive environments starting with the list developed earlier for landfills, and tailoring this to meet sewage treatment facility related environments.

TABLE 10-32
POTENTIALLY SENSITIVE ENVIRONMENTS
AND OTHER FACTORS LIMITING SEWAGE TREATMENT
FACILITY DEVELOPMENT

Wetlands
Flood Plains
Threatened & Endangered Habitats
Scenic Areas
Natural Screening
Development (Exist & Committed)
Areas of Historic Importance
Areas of Architectural Importance
Areas of Paleontological Importance
Areas of Archaeological Importance
Natural Buffering
Close to Major Highway
Highway Restrictions
Close to Centroid
Available Land

The second matrix can now be developed as shown in Table 10-33.

TABLE 10-33:
MATRIX OF ENVIRONMENTAL IMPACTS OF SEWAGE TREATMENT FACILITY
AND POTENTIAL SITING CRITERIA

**Environmental Impacts**

| Sensitive environments | Use of land for building | Highway congestion / Sludge haul | Highway noise / Sludge haul | Accidents / Sludge haul | Noise / Excavation | Noise / Fill | Noise / Building | Noise / Sludge haul | Air pollution / Treatment | Air pollution / Emissions | Air pollution / Sludge haul | Economic loss / Treatment | Reduction of scenic quality / Building |
|---|---|---|---|---|---|---|---|---|---|---|---|---|---|
| Wetlands | X | | | | | | | | | | | | |
| Flood Plains | X | | | | | | | | | | | | |
| Threatened and endangered species habitat | | | | | X | X | X | | X | X | X | | |
| Scenic Areas | X | | | | | | | | | | | | |
| Natural Screening | | | | | | | | | | | | | X |
| Development (Existing & Committed) | | | | | X | X | X | X | X | X | X | X | |
| Areas of historic importance | X | | | | | | | | | | | | X |
| Areas of architectural importance | X | | | | | | | | | | | | X |
| Areas of paleontological importance | X | | | | | | | | | | | | |
| Areas of archaeological importance | X | | | | | | | | | | | | |
| Natural Buffering | | | | | X | | | | | | | | |
| Proximity to major highway | | | | | | | | | | | | | X* |
| Highway Restrictions | | X | | | | | | | | | | | |
| Proximity to centroid | | | X | X | | | | | | | | | |
| Availability of land | X | | | | | | | | | | | | |

174

## TABLE 10-34
## SELECTION CRITERIA FOR SEWAGE TREATMENT FACILITY DEVELOPMENT

Wetlands
Flood Plains
Threatened & Endangered Habitats
Scenic Areas
Natural Screening
Development (Exist & Committed)
Areas of Historic Importance
Areas of Architectural Importance
Areas of Paleontological Importance
Areas of Archaeological Importance
Natural Buffering
Close to Major Highway
Highway Restrictions
Close to Centroid
Available Land

From the matrix shown in Table 10-33, we can develop the final list of selection criteria as shown in Table 10-34.

Now we need to divide this list into regional and local criteria.

## TABLE 10-35
## REGIONAL SITING CRITERIA
## FOR SEWAGE TREATMENT FACILITY DEVELOPMENT

Wetlands
Flood Plains *
Proximity to Major Highways **

 * A minimum distance of 500 feet is suggested.

** A maximum distance of 500 feet is suggested.

The local siting criteria are presented in Table 10-36.

## TABLE 10-36
## LOCAL SITING CRITERIA
## FOR SEWAGE TREATMENT FACILITY DEVELOPMENT

Threatened & Endangered Habitats
Scenic Areas
Natural Screening
Development (Exist & Committed)
Areas of Historic Importance
Areas of Architectural Importance
Areas of Paleontological Importance
Areas of Archaeological Importance
Natural Buffering
Close to Major Highway
Highway Restrictions
Close to Centroid
Available Land


## RECYCLING FACILITIES

The construction and operational phases of recycling facility development are identified in Table 10-37.


## TABLE 10-37
## CONSTRUCTION AND OPERATIONAL PHASES
## OF RECYCLING FACILITY DEVELOPMENT

Excavation
Fill
Building
Haul to Recycle
Queue at Facility
Material Deposition
Separation
Storage
Air Emissions
Residual Haul

The environmental impacts associated with these phases of construction and operation are as follows:

TABLE 10-38
ENVIRONMENTAL IMPACTS ASSOCIATED
WITH RECYCLING FACILITY DEVELOPMENT

Use of Land
Highway Congestion
Highway Noise
Accidents
Noise
Air Pollution
Economic Loss
Reduction of Scenic Quality

Adding these two lists together as in previous examples produces a matrix of construction and operational phases of recycling facility development versus environmental risks.

# TABLE 10-39:
## MATRIX OF CONSTRUCTION AND OPERATIONAL PHASES OF RECYCLING FACILITY DEVELOPMENT VERSUS ENVIRONMENTAL IMPACTS

**Environmental Impacts**

| Construction and Operational Phases | Use of land | Highway congestion | Highway noise | Accidents | Noise | Air Pollution | Economic loss | Reduction of scenic quality |
|---|---|---|---|---|---|---|---|---|
| Excavation | | | | | X | | | |
| Fill | | | | | X | | | |
| Building | X | | | | X | | X | X |
| Haul to recycle | | X | X | X | | | | |
| Queue at facility | | | | | X | | | |
| Material deposition | | | | | X | X | | |
| Separation | | | | | X | X | | |
| Storage | | | | | | | | |
| Air emissions | | | | | | X | | |
| Residual haul | | X | X | X | X | | | |

## TABLE 10-40
### POTENTIAL ENVIRONMENTAL IMPACTS
### OF RECYCLE FACILITY DEVELOPMENT

Use of Land for Building
Highway Congestion/Haul to Recycle
Highway Congestion/Residual Haul
Highway Noise/Haul to Recycle
Highway Noise/Residual Haul
Accident/Haul to Recycle
Accident/Residual Haul
Noise/Excavation
Noise/Fill
Noise/Building
Noise/Material Deposition
Noise/Separation
Air Pollution/Material Deposition
Air Pollution/Separation
Air Pollution/Air Emissions
Economic Loss/Building
Reduction of Scenic Quality/Building

With this first matrix completed, we can identify potential environmental impacts which may occur as a result of one or more aspects of recycle facility development.

Which environments provide the best protection against these impacts? Which environments are most sensitive to these types of impacts? These are the questions to be answered in the development of the second matrix.

As before the next step is to identify the potentially sensitive environments starting with the list developed earlier for landfills, and tailoring this to meet recycling facility related environments.

## TABLE 10-41
## POTENTIALLY SENSITIVE ENVIRONMENTS
## AND OTHER FACTORS LIMITING RECYCLING FACILITY DEVELOPMENT

Wetlands
Flood Plains
Threatened & Endangered Habitats
Scenic Areas
Natural Screening
Development (Exist & Committed)
Areas of Historic Importance
Areas of Architectural Importance
Areas of Paleontological Importance
Areas of Archaeological Importance
Natural Buffering
Close to Major Highway
Highway Restrictions
Close to Centroid
Available Land

The second matrix can now be developed as shown in Table 10-42.

## TABLE 10-42:
## MATRIX OF ENVIRONMENTAL IMPACTS OF RECYCLING FACILITIES AND POTENTIAL SITING CRITERIA

**Environmental Impacts of Recycling Facility Development**

| Sensitive Environments | Use of land for building | Highway congestion / Residual haul | Highway congestion / Haul | Highway noise / Residual haul | Highway noise / Haul | Accidents / Residual haul | Accidents / Haul | Noise / Residual haul | Noise / Excavation | Noise / Fill | Noise / Building | Noise / Material deposit | Air pollution / Material deposition | Air pollution / Separation | Air pollution / Separation | Economic loss / Emissions | Reduction of scenic quality / Building | Reduction of scenic loss / Building |
|---|---|---|---|---|---|---|---|---|---|---|---|---|---|---|---|---|---|---|
| Wetlands | X | | | | | | | | | | | | | | | | | |
| Flood Plains | X | | | | | | | | | | | | | | | | | |
| Threatened and endangered species habitat | X | | | | | | | X | X | X | X | X | X | X | X | | | |
| Scenic Areas | | | | | | | | | | | | | | | | | | |
| Natural Screening | | | | | | | | | | | | | | | | | | X |
| Development (Existing & Comm.) | | | | | | X | X | X | X | X | X | X | X | X | X | X | | |
| Areas of historic importance | X | | | | | | | | | | | | | | | | | X |
| Areas of architectural importance | X | | | | | | | | | | | | | | | | | X |
| Areas of paleontological importance | X | | | | | | | | | | | | | | | | | |
| Areas of archaeological importance | X | | | | | | | | | | | | | | | | | |
| Natural Buffering | | | | | | | | X | X | X | X | X | | | | | | |
| Proximity to major highway | | | | | | | | | | | | | | | | | | |
| Highway Restrictions | | X | X | | | | | | | | | | | | | | | |
| Proximity to centroid | | X | X | X | X | | | | | | | | | | | | | |
| Availability of land | | | | | | | | | | | | | | | | | | |

181

TABLE 10-43
SELECTION CRITERIA FOR RECYCLING FACILITY DEVELOPMENT

Wetlands
Flood Plains
Threatened & Endangered Habitats
Scenic Areas
Natural Screening
Development (Exist & Committed)
Areas of Historic Importance
Areas of Architectural Importance
Areas of Paleontological Importance
Areas of Archaeological Importance
Natural Buffering
Close to Major Highway
Highway Restrictions
Close to Centroid
Available Land

From the matrix shown in Table 10-42, we can develop the final list of selection criteria as shown in Table 10-43.

Now we need to divide this list into regional and local criteria.

TABLE 10-44
REGIONAL SITING CRITERIA
FOR RECYCLING FACILITY DEVELOPMENT

Wetlands
Flood Plains *
Proximity to Major Highways **

\* A minimum distance of 500 feet is suggested.

\*\* A maximum distance of 500 feet is suggested.

The local siting criteria are presented in Table 10-45.

## TABLE 10-45
## LOCAL SITING CRITERIA
## FOR RECYCLING FACILITY DEVELOPMENT

Threatened & Endangered Habitats
Scenic Areas
Natural Screening
Development (Exist & Committed)
Areas of Historic Importance
Areas of Architectural Importance
Areas of Paleontological Importance
Areas of Archaeological Importance
Natural Buffering
Close to Major Highway
Highway Restrictions
Close to Centroid
Available Land

# 11

# The Way Forward

## THE ENVIRONMENTAL COALITION

Concern to safeguard the quality of our environment is increasing with each passing year. As a result, every proposal to develop a landfill is greeted with outrage as though landfill disposal is part of the problem rather than part of the solution.

It is important to build a partnership between those who would safeguard the environment by resisting all development, and those who would plan intelligently for future growth.

Landfill planning is common sense. There is no disregard for the environment in this statement. As a nation we simply produce more solid waste, in greater varieties than we can ever handle by alternative technologies. In addition, every alternative technology generates its own unique category of residuals, which in the absence of more alternative technology, goes to landfill.

Other LULUs referred to earlier are vitally necessary to safeguard the environment and manage the remaining natural resources. The need for transfer stations is unassailable. As urban development continues to spread, solid waste treatment or disposal capability will continue to be pushed further and further from the centers of waste generation. While solid waste is efficiently collected in trucks designs for house to house collection, it makes no sense for these same trucks to haul the waste long distances to remote sites. Not only is this a wasteful utilization of energy resources because packer trucks are not designed for long hauls, but it takes packers out of the collection service necessitating additional expenditures on more packers to service collection routes. Transfer stations fill the gap in the service system. They provide a nearby transfer system for packers from which they can return to collection most efficiently while transfer trailers designed for long hauls can take over to carry solid waste to its final destination.

Incineration is necessary if we are to recover energy resources from solid waste. Without detracting from the necessity to recover material resources from solid waste, it must be recognized that many materials are simply not recoverable. Very often materials cannot be recovered because there is no market for the recoverable item. Another category of non-recoverable solid waste includes items which are so contaminated that recovery itself would require the use of additional cleaning agents which in their turn would only generate more waste. Complex non-homogeneous items of waste such as mattresses, electrical equipment, refrigerators, appliances, all fall into the category of non recoverable waste for which final disposal can be landfill or incineration. In the 1990's, communities will turn to incineration in increasing numbers.

Communities are already turning to composting of yard waste to relieve the burden upon sanitary landfills. It surely makes conservation and economic sense to recover valuable soil nutrient and conditioning resources, rather than bury them in landfills.

Sewage treatment needs grow with population increase, and the growth in urban development.

Recycling centers are needed to serve as collection points for a wide variety of recyclable materials which are currently discarded through lack of knowledge or convenient collection.

Hazardous waste must be handled and treated sensibly if we are to avoid the superfund sites of tomorrow.

Beyond waste management, there are LULUs which are also critical to future growth and development. Quarries, asphalt plants, and bulk handling plants to name just a few, all need careful planning and all are needed. We simply must learn to manage the siting process.

The most informed groups very often are those commonly considered to be the opposition. The environmental action groups very often find themselves in the foreground of the opposition not quite knowing how they got there. A coalition is possible between developer and environmental action group to the extent that these organizations can become the projects most ardent supporters.

To achieve this miracle, the siting process must be appropriate and thorough, and it must be done right.

## CO-OPTING PUBLIC SUPPORT

Co-opting public support requires persistence, patience and careful planning. When environmental action groups can be successfully included, the process is significantly easier because the public at large often looks to these groups for guidance. If the siting process is clearly supportable, environmentally well informed individuals will literally work to gain approval as ardently as opposing any loss of wetland habitat or any cutting of old growth timber.

However, support from environmental groups is not essential to success. Public support is a necessity. It is important here to identify who the public is in this context. Using landfill siting as an examples, we need to remember the two step siting process; regional and local. During the regional siting process, the public consists of active members of the community interested in supporting a well planned siting process. The public is important in establishing a sensible siting procedure. During the local siting process, a hierarchy of specific sites emerges, and a final site is selected. Many planners fail to perceive that "the public" changes by definition at this point, and the potential exists for an active and vociferously anti-landfill public to emerge at this stage.

When a site is chosen, it is important to co-opt public support immediately. This can only be achieved by door-to-door visits. However difficult this may sound, and however attractive other alternatives may appear, ignore all other options. Public meetings, town meetings, church socials, cocktail parties can also be tried later in the process if required, but door-to-door meetings with those in the immediate vicinity should not be avoided.

In the view of the writer "in the immediate vicinity" should mean within 1 mile radius of the landfill property including a buffer zone, not one mile from the edge of where the landfilling activity begins.

Make a list of all names, shops, offices, farms or public buildings because all members of the public should be included in this process. Telephone every establishment, and make an appointment. Explain that you want to hear their views. Remember the first rule is to be a good listener. Remember that a great deal of public opposition arise from frustration at always being told what to do. The investigator is there primarily to receive information, to find out what the public feels about a proposed landfill.

Make notes. Clearly record views. Do not trust that expressions of concern will be remembered after the interview is over. Write them down. Remember to promise to get back to a questioner with more information where necessary and remember to keep the promise.

At the end of this process provide information where possible.

Be sure that the information is factual and accurate.

BUILDING PUBLIC TRUST

A good deal of the preceding discussion on co-opting public support has centered on building public trust. The public will not support you unless you are trusted.

Attempts to "sell" the landfill will rarely build trust because it will be clear that all you want is their vote. Center your effort on being committed to the public getting all the information available to make an informed decision, and the trust will follow.

Things that get in the way of building public trust are:

1.  Past performance of landfills either locally or nationally.

2.  Past performance of the group promoting the landfill either as individuals or the group as a whole.

As far as landfill history is concerned, the first of these problems should not be at all surprising. For centuries, landfills in many areas have been little more than holes in the ground with little or no control of the material being burned or buried in the hole.

As a result of this kind of history, so called landfills have been responsible for polluting groundwater and literally laying waste hundreds of acres of valuable land which now appear to be consigned to a real estate limbo where they have little or no value except as enormous liability. This is not a neighborhood land use which the local population could be expected to view without concern.

Only in the last ten years has it become clear that leachate must be controlled before it leaves the landfill.

Only by describing clearly and concisely how this design differs from past practices will it be possible to build public trust.

Similarly, inadequate and poorly maintained cover design have in the past increased the risk of leachate contamination of public water supplies. An explanation of how a modern landfill cap is designed is essential together with details of how post closure care provides for on-going maintenance to prevent erosion.

Every effort must be made to show how modern landfills differ from those historic problems which still haunt the industry. Dealing with the past performances of individuals or the group promoting the landfill is beyond the scope of this book. It should be sufficient here to point out that these things do have an impact upon public perceptions and public trust.

There are several new developments in dealing with LULUs in general, and landfills in particular, which deserve special attention here because they may help significantly in dealing with the whole question of public trust. These developments may usefully be discussed under the general headings of:

> Contractual Agreements
> Quality Assurance Programs
> Compensation Programs

## CONTRACTUAL AGREEMENTS

### Duration Limits

One of the major complaints heard at almost every anti-landfill meeting is that once a landfill is approved for say 20 years, the nearby community has no control over its actual duration. There are countless examples of landfill developments which have started as 20 year operations which have returned to regulatory agencies again and again for expansion. Naturally, the local community has just cause to resent this process. A community which makes an agreement to permit a landfill for a fixed period of time has a right to expect that agreement to be honored. It is time that landfill developers agreed to put landfill duration in a contract, subject to ratification by the courts if necessary.

TABLE 11-1

ITEMS OF LANDFILL DEVELOPMENT NECESSARY
FOR OPERATION REGARDLESS OF SIZE AND DURATION

Administration Building
General Offices
Washrooms
Locker Rooms
Gate House
Scales
Fence
Security
Garage
Vehicle Maintenance
Permanent Road

The lack of willingness on the part of landfill developers to commit to a fixed period of operation has been a serious problem in the past.

Similarly, the lack of willingness to look at shorter planning periods is also a problem. Commonly, the argument against a shorter period is "savings in scale". However all too often this phrase is not fully understood and it amounts to short-hand for a general unwillingness to look at the options.

Typically savings in scale simply implies that the landfill must be as large as possible and operate for as long as possible in order to reap the benefits of all the fixed development necessary in order to operate a landfill efficiently. These fixed development items are identified in Table 11-1.

Very often, the larger the planning period, the more grand and costly some of these facilities.

In the past, these fixed costs were a significant proportion of the total cost of landfill development because the capital cost of the land itself was the only remaining development cost to be considered. In recent years, the equation has changed significantly. While the fixed developments have stayed the same. The preparation of the land, monitoring, closure, and post closure care requirements have altered out of all proportion to the fixed costs. A short list of current landfill development requirements by no means intended to be complete will suffice to illustrate the point:

TABLE 11-2

ITEMS OF LANDFILL DEVELOPMENT
DIRECTLY RELATED TO LANDFILL SIZE

---

Clay Recompaction
Quality Control of Base
Flexible Liner (if required)
Leachate Collection Pipes
Gas Collection Pipes
Vegetative Cap
Groundwater Monitoring
Gas Monitoring
Leachate Withdrawal System
Leachate Storage
Gas Extraction System
Gas Storage
Erosion Control
Drainage Control
Erosion Monitoring

---

In a recent study, for the Village of Bartlett, Illinois, the writer has shown that in the case of a proposed balefill development the costs of developing a single 20 year site were less than the cost of developing four 5 year sites by only about 4.5% of the total. The writer found that it would have been easier to find four small sites close to the center of waste generation than one large site. The energy savings over a twenty year period amounted to over $34 million or 20% of the transport costs to the single site option.

A willingness to look at shorter planning periods than 20 years has significant advantages not only in increasing the number of options, but also in improving the prospects of approval. A landfill designed to operate for only 5 years may be an attractive offer to a community if it comes with an agreement that after 5 years, it will be definitely closed with no prospect of continued expansion.

Tonnage Limits

Next to continuously expanding landfills, the public at large has a particular problem with the landfill which starts small and grows rapidly to a point where the traffic impact became intolerable. A contractual agreement which limits the specific daily tonnage permitted is needed to protect the public from major increases in daily traffic. As in the case of the duration limit, there are siting advantages in a willingness to make such agreements with the local community.

# QUALITY ASSURANCE PROGRAMS

## Liner Integrity

In modern landfill design, no quality assurance program is more important than that which assures the integrity of the liner. Very often this program is a requirements of the regulatory agency. Where it is not a state requirement, the landfill developer should carry out the procedure in any event as insurance against future problems, but also as public assurance.

It is not yet fully appreciated by those in the waste management industry that this kind of quality assurance program is as important to the public as it is to the owner and operator of the facility.

Prospective opponents of a facility who feel that their water supply is threatened should be encouraged to see information on the liner integrity program. Prospective landfill developers should be encouraged to share the details of the program with all members of the public. There is no more effective way to get members of the public to see reason than to share the details with them.

If possible, where conditions at the site permit, efforts should be made to transport members of the public out to the site area where recompaction or preparations for laying a liner, or seam testing are taking place. Allow questions to be asked. Answer questions where possible, get back to questioners where more detail is necessary. None of this information sharing is as dangerous as some landfill developers appear to think. The goodwill that information sharing brings is reason enough to do it. However, there is a much more serious purpose and that is that quality assurance programs are not for the developer alone, they are for the public which feels threatened by the development.

## Progress Reports

Landfill development is difficult to manage. Unlike a major building construction project, it is difficult to see results until close to the end of the project.

The public in the vicinity of new landfill development find that the process is bewildering because little or no information is shared about the progress of the development. With the atmosphere of suspicion which often surrounds this kind of development, this level of public ignorance breeds more dissent, rumor, and potential court action.

The remedy is to share the information. Share progress reports with the local community officials, and with members of the public within one mile radius of the site. Initiate a monthly or a quarterly newsletter. Get out all the information on what is going on at the site, including inspection reports, monitoring results, pump failures, and even illegal loads turned away.

This is not secret information, it is information of vital interest to all those living in the area. Given the nature of the resistance to siting, the willingness to share this information will significantly improve the prospects of approval.

## COMPENSATION MEASURES

### Host Communities

When a private developer of a regional pollution control facility purchases land in a host community the increased valuation of the property adds to the tax base in the community. When a public developer, such as a county solid waste management implementation agency purchases property, the tax revenues are effectively lost to the host community.

In many cases where the proposed development involves substantial local employment the benefits to the host community outweigh the loss of tax revenues. However, in the case of a regional solid waste management facility, the increase in employment required by the facility is rarely greater than 20-50 employees, even in the case of a large waste-to-energy operation. In the case of a large landfill the number of employees is rarely more than 10-25.

Thus, it is common practice particularly in the case of waste-to-energy facilities for a public agency to negotiate a host community fee to compensate the local government for the loss in revenue.

A great many of the potential impacts upon a host community need no compensation because they can be mitigated both by design and operation, and also by setting up reporting procedures which provide the host community with hard evidence that the mitigation is working.

However, some impacts are not easily handled by mitigation and these are more amenable to compensation. In the following pages compensation techniques will be discussed in relation to each specific type of facility.

### Sanitary Landfills

From the viewpoint of a host community, a new landfill does provide some small amount of direct local employment, in addition to some indirect employment created by an increased need for heavy equipment maintenance and landscaping needs. The term indirect employment is used here to describe the increased employment needs of existing businesses in the area which might arise from providing goods and services to the facility. Heavy construction equipment, trucks, building supplies, gravel, sand, PVC pipe and flexible membrane liner are major items of capital equipment, some of which will be supplied from local sources. Ultimately, in terms of landfill gas supply, it is possible that the indirect employment benefits of the landfill could be considerable if this new source of energy attracts new industry to the area.

192

On the negative side from the perspective of the host community, packer truck traffic on local roads will increase road maintenance costs and possible contingency uses of the local fire department may increase fire department costs. Leachate treatment will also increase wastewater loads at the publicly operated treatment works (POTWs).

It is common practice in a waste-to-energy operation to include a negotiated assessment of these costs in the host community fee.

A host community fee is an appropriate vehicle by which to compensate a host community for the risks associated with the development of any solid waste management systems. Water supply and sanitary treatment requirements at landfills are typically handled by on-site systems but in the event they are met by public services, an assessment of these cost impacts should also be included in the host community fee.

The potential environmental impacts are summarized in Table 10-2. Those factors amenable to compensation are the potential impacts upon community water supplies, reductions in tax base due to the perceived potential for loss of property values, and perceptions that the local community will be considered the dumping ground for other peoples garbage. This list is not intended to be all inclusive. There may be special circumstances which add or subtract from this list. The point is that no matter how unclear these impacts may be they are all viable items for negotiated settlement.

For example, in a later section of this chapter a real estate value assurance program is described which will monitor house prices and reimburse home owners for loss in value. This same technique can be used to reimburse a host community for any collective loss in tax revenues.

It should be noted here that, contrary to popular belief, there is almost no evidence that property values are negatively impacted by sanitary landfill development. Nevertheless, while the community perception is that the risk is real, a value assurance program can reduce the concern. The complete system will be described in detail later under private citizen compensation but basically the value assurance scheme involves the identification of properties of equivalent market value outside the field of influence of the landfill. The prices of these surrogate properties can be monitored annually for changes in value compared to properties in the area of the landfill. Any reductions in the value of property which would impact real estate taxes can be compensated by the implementing agency.

Similarly, a water quality assurance agreement can be established between an implementation agency and a host community which would guarantee zero impact upon a municipal water supply due to the landfill operation. If the source of municipal water is polluted by the landfill the agency will replace it with an equivalent water source.

What is required here is a three part program:

1.  Assurance that water quality is being maintained by a willingness to share groundwater monitoring data.

2.  Compensation if water quality falls below a certain level.

3.  Replacement of supply if renovation of the systems proves impractical.

The willingness of the landfill operator to enter into such an agreement is derived from the assurance that investment in groundwater monitoring systems assures the capability to detect any leachate losses from the landfill long before there is a groundwater problem at the local well.

Even such an unclear impact as the concern that the community will come to be considered a dumping ground is capable of settlement by negotiation. The problem can be approached initially by identifying any tangible features of landfill siting which contribute to the overall concern. These features may include the name of the operation, the potentially poor condition of the surrounding roads, and the view of the site. Most aspects of these concerns are capable of mitigation by siting, design, or by a clear definition of operating policy.

When those elements of the dumping ground syndrome, which can be mitigated, are peeled away from the overall concern, there remains a small element of impact that is a function of the community perception. This can be the subject of compensation in the form of an adjustment to the host community fee.

It should be noted that compensation can take a variety of different forms. While the host community fee is the most common it is not unusual for specific services to be part of the agreement with the local government. For example, part of the compensation package could include reduction or elimination of disposal costs for the host community.

Waste-to-Energy Facility

A waste-to-energy facility provides substantial direct employment during an approximate two year construction period and some additional employment during operation. Although fewer items of heavy equipment are involved than for a landfill, there is likely to be some need for maintenance services and supplies from within the local community. Electrical, mechanical and computer equipment will typically generate indirect local employment and during construction a wide variety of building supplies will be required and local suppliers will have obvious advantages in the bidding process.

Ultimately, the waste-to-energy facility will be a local source of low cost energy with the potential to attract high energy use industries into the community. Finally, modern waste-to-energy technology is sufficiently new that many host communities literally bask in the perception that they are in the vanguard of solid waste management. The potential environmental impacts which must be considered in the compensation agreement are summarized in Table 10-13.

Packer truck traffic on local roads will increase road maintenance costs in addition to increased traffic during construction. There is a small contingent risk that support might be needed from the local fire department. Washwater flow to the local POTW will increase BOD and suspended solids loading, although this will be directly compensated by increased sewer surcharges.

These are all tangible items which are typically included in a negotiated host community fee. The risk items that are less easy to quantify are tax base reduction due to perceived potential loss of property value and the notion that the community will be considered the dumping ground. Property value reductions can be handled by a value assurance program as described earlier in the case of landfill compensation. The dumping ground syndrome is weaker in the case of a waste-to-energy operation in spite of the fact that the facility is more prominent because it is offset to some extent by the forefront-of-new-technology-syndrome. However, as in the landfill example, these perceived risks can be included in the negotiations over the host community fee.

The perceived risk of major quantities of air pollutant emissions is principally a mitigation concern. However, once the details of reporting procedures have been worked out the fact remains that there may still be a perceived risk that something could go wrong to drastically change the emissions. While the designer and the agency may have complete faith in the emission controls and in the back-up systems, what the community needs at this point is a mechanism by which they can be compensated if anything goes wrong. It seems inappropriate that this should be part of the host community fee since the obligation of the agency and the designer is to reduce this risk to an absolute minimum. It is important to note that the agency also has an obligation to educate the host community in the various ways in which the emissions are controlled. Experience indicates that in the final analysis, some communities still have misgivings about emissions no matter how complete the educational program. It is therefore proposed that a performance agreement be established with the host community separate from the community fee. In the event that emissions fall below prescribed levels and for every day the emissions are out of compliance, a negotiated penalty should be paid to the community in the form of a specific monetary penalty or reduced fees for solid waste collection services.

A potential environmental impact frequently associated with waste-to-energy facilities derives from the possibility that the ash generated by these facilities may have some of the characteristics of a          hazardous waste. Aside from the risks associated with possible air emissions from ash handling activities, the risk is principally concerned with the after effects of landfilling the ash. The risk of polluting the community water supply due to ash disposal is essentially part of the previously discussed agreement with the landfill operation. Whether the risk is due to ash or solid waste disposal does not impact the compensation technique.

Transfer Stations

Transfer stations provide a simple means by which the host community can substantially reduce haul costs when the disposal facility is remotely located from the community. The potential environmental impacts are summarized in Table 10-4.

There are potential impacts of increased road maintenance costs, potential calls on the local fire department, water supply and sewer service costs. These factors are easily quantified and made part of the host community fee.

Less tangible is the risk of tax base reduction due to the perceived potential for reduction of property values and this can be handled as identified earlier with a value assurance program.

Composting

Leaf waste and other yard waste take up a substantial volume of scarce landfill capacity. This waste material is not only inoffensive compared to residential solid waste, but if properly managed can be a valuable resource as a soil conditioner returning valuable humus to the natural environment.

The advantage to the host community is a ready supply of composted mulch which can be used on municipal landscape projects.

A large scale operation has the advantage of being able to justify expenditure on heavy equipment capable of accelerating the natural process of decomposition referred to as composting. The only risks involved are those related to a breakdown in the system, since normal operation should be entirely acceptable to the host community. A host community fee seems entirely inappropriate to establishing a composting operation since the impacts to the community appear to be largely positive. Risks of breakdown should be handled with a performance agreement. Such a breakdown might include the midnight dumping of a hazardous waste or municipal refuse on to the composting pile. Typically, contingency plans are prepared in order to deal with this problem. However, we must suppose that the risk also includes the breakdown in the contingency plan to the extent that there is a hazardous discharge from the composting facility and a risk of groundwater contamination. Obviously, every precaution must be taken to prevent this from happening, but nevertheless, the community should have the assurance that if such a breakdown occurred, compensation is available. This assurance can be negotiated as part of a performance agreement with the host community.

A three part program is proposed:

1.  Visual policing on a daily basis to detect any promiscuous dumping and a daily log of this inspection. A monthly summary of this log should be provided to the host community.

2.  If a promiscuous dumping event is detected an automatic penalty will be paid to the host community for every day the waste material is allowed to remain on the composting facility. The amount of this penalty to be the subject of negotiation with the host community.

3. A groundwater monitoring system is required capable of detecting any leachate movement from the composting operation into the subsoil. Appropriate parameters to be analysed must be agreed beforehand in addition to maximum levels of those parameters above which the penalty will be paid.

It is proposed that the groundwater be monitored on an semi-annual basis (since the risk of promiscuous dumping is very small) and that a copy of this report be provided to the host community.

Naturally, if there is any increase in previously identified parameters in the groundwater the community must have the assurance based on a written agreement that the water quality will be returned to the condition which existed prior to establishing the operation or an alternative supply will be provided at no cost to the host community.

Private Citizens

Sanitary Landfills

Most of the potential environmental impacts associated with landfill development have been discussed and dealt with either by mitigation or a combination of mitigation and reporting requirements.

Groundwater pollution is also handled by mitigation and by an established procedure for reporting groundwater monitoring results. What happens if all of these protection measures fail to fully protect the system?

The designer looking at this problem will consider that the groundwater monitoring program will provide an early warning long before the wells are threatened. If the system works as designed, and if the wells are properly sited and monitored, then the protection systems should be adequate. However, the public citizen is dealing with the potential loss of a valuable resource and he or she has every right to ask - what if the system fails?

Clearly, a contingency plan is required and this should be agreed to early in the process. The plan should be clear and explicit regarding what the implementation agency will do and when.

The appropriate parameters of water pollution must be agreed to early since the program cannot be expected to protect private wells from all forms of pollution but only those caused by leachate from the landfill.

A radius of 1,000 feet is proposed as the area of influence. Clearly, residents beyond this area will feel that they are also at risk, but they will also recognize that wells nearer the facility will show signs of pollution first and if properly tested will provide a safeguard to their own sytems.

All private potable water supply wells within 1,000 feet of the perimeter of the fill area should be initially tested for the complete range of priority pollutants currently stipulated by the appropriate regulatory agency. This initial full scale analysis is necessary in order to determine if the existing wells already have characteristics which could generate false positive results in later testing. If there are no private wells within 1,000 feet the range should be extended by 100 foot increments until at least 10 wells are identified. The maximum radius of influence proposed is 2,000 feet.

If there are no wells within 2,000 feet it is suggested that the implementing agency bore their own indicator wells and provide concerned citizens with the relevant data on a quarterly basis.

Ideally, private wells should be found which are representative of each quadrant surrounding the landfill which is the reason why the radius should be extended further as necessary. Initial tests will provide background levels to compare with those to be taken during the operation. The advantage of having a number of wells is that they will be corroborative if a problem occurs and if contamination due to errors in sampling are suspected.

There are several technical issues which must be identified in the agreement in order to avoid premature reaction by the implementing agency. The first issue is concerned with choosing the right indicators. It is well established that the first indicators of pollution are large molecule anions. This is because the anionic exchange capacity of soil is generally low and the large molecule anions are the least amenable to adsorption on the soil matrix. Thus chlorides, sulphides and carbonates are the first indicators of leachate contamination. However, these salts exist in large concentrations in many private wells and they are the principal source of hardness. They also vary seasonally and annually.

Thus, at a minimum, the initial private well monitoring for background conditions should cover at least all four seasons. Since initial construction will extend over at least 10-12 months, this is not particularly onerous, although it does require forward planning. The aim is to establish a sufficient number of tests to identify within a broad range the extent to which each of the indicator parameters will vary.

When operation of the facility begins, it is suggested that well owners take their own test samples and have them collected by the implementing agency or a laboratory of their choice. One small problem with this approach is that the test procedure must be in conformance with a standard practice, often referred to as a protocol, and homeowner sampling may not provide sufficient reliability for analytical purposes. However, the advantage of placing this responsibility in the hands of the well owner outweighs the difficulties and a simple training program is proposed to ensure compliance with the sampling and sample storage protocols.

One final issue must be discussed and that is how to deal with anomalous results. As a practical matter, groundwater quality varies for many reasons and there must be an agreement which will permit the agency to return and resample before the water quality guarantee provisions go into effect. If after resampling and analysis a private well shows elevated levels of any two of the indicator parameters then the compensation provisions should be initiated.

Homeowners with private potable water supply wells can be compensated in a number of ways. The first option, if the level of contamination is low, is to return the groundwater supply to its original condition. This can be achieved by placing a barrier between the source of pollution and the well or wells. Alternatively, a deeper aquifer can be drilled which would be capable of supplying water to a number of private citizens within a limited area. Connection to the public water supply is a third alternative.

This type of compensation can be costly, but it should be remembered that a modern facility is extremely unlikely to cause the problem in the first place. We are dealing with "what if" concerns. The private citizen not only needs assurance that the worst will not happen, but assurance that if it does happen a written agreement is in place which guarantees correction. The implementing agency needs assurance that the prospect of having to compensate for loss of water quality are extremely limited and that can be achieved by stringent application of the many quality control features of modern landfill practice.

The second and even greater potential environmental impact which private citizens perceive as being synonymous with nearby landfill development is loss of property value.

A great many studies have been completed in various parts of the country to evaluate the impact of regional waste treatment facilities upon real estate values. In most cases, the conclusion is that there is either no impact or the rate of increase is marginally impacted.

Nevertheless, it must be recognized that investment in a home is for many private citizens their single major life investment and any development which impacts the increasing value of that home is bitterly resented.


In a 1978 report[2] from the U.S. Department of Agriculture, Division of Economics, Statistics and Cooperative Services, the authors acknowledge the importance of dealing with the question of declining property values:

> "To deal effectively with citizen opposition to landfill siting, State and local officials must concern themselves with the effect the newly established landfill will have on property values. This report gives the findings of a survey of resident attitudes

toward landfills in four communities in Illinois, Indiana, and Wisconsin. The study found that factors associated with landfills which tended to cause declining property values included water pollution, blowing papers from the site, rodents and insects, and odors. Any plan regarding disposal sites needs to address such problems."

---

[2] "Attitudes of Nearby Residents Toward Establishing Sanitary Landfills." U.S.D.A. Report ESCS-03, January 1978

"Of those respondents who felt their property values were affected by the landfills, most believed that these values declined. Those holding such an opinion were more likely to take action to express their feelings than were those who felt the disposal site did not affect their property values. In addition, the greater the value that a respondent placed on his property, the more likely he was to oppose the establishment of the disposal facility, and take action against it."

"About 70 percent of the residents in the survey areas felt that proposed landfills should be established at least one mile from the residences. Those preferring a greater distance held a less favorable opinion of its establishment. The closer a respondent lived to a site, the more apt he was to be unfavorable toward it and to express his opinions about it."

"A number of promises made by State and local officials did little to encourage public acceptance of landfills. Promises relating to improved collection and disposalservices, free access to the site, screening of the site, and development of the site as a future park or recreation area had little influence on the opinions of either the proponents or opponents. Concerns about open burning at the site and excessive noise had only moderate influence on such attitudes."

We believe that a property value assurance program can be put in place which will provide a guarantee that if property is sold for less than the fair market price the difference will be paid to the homeowner in full.

This proposal depends upon a real estate appraisal technique referred to as equivalent market value. Stated simply, all property within a certain range of the facility (a range of half a mile is suggested) will be appraised by a real estate appraiser approved by both parties. The same appraiser is then required to identify homes of equivalent market value in areas where growth, access and other features of these surrogate communities are as close as possible to the areas in question. A comparison of market value is required on an annual basis in order to keep abreast of changes in valuation.

When a house in the landfill area is sold, the price is compared with its surrogate. If the price of the house near the landfill is lower than that of the surrogate the difference is paid to the owner to compensate for loss in value.

Clearly, this system is only an approximation since there can be no absolutely equivalent house where the location and all the many parameters which contribute to value are identical. However, methods of comparing equivalent homes are commonplace procedures in the real estate appraisal and the practice is not without sophistication. The accuracy of the system can be improved significantly the greater the number of equivalent homes identified. Naturally, this increases the cost of real estate appraisal services, but the agency may choose to err on the side of increased accuracy where the prospect of local support is at stake.

## Waste-to-Energy Facilities

The majority of potential environmental impacts to private citizens associated with waste-to-energy facilities have been discussed earlier under mitigation. The only risk remaining which requires further discussion here under compensation is that of loss in property value.

A property value assurance program is proposed identical to that discussed earlier under Sanitary Landfills. The major difference will be in identifying the area of homes to be considered part of the program.

## Transfer Stations

Claims have been made that transfer station location has an impact on local property values. A similar program of property value assurance is proposed in order to address this possible concern.

## Composting

A property value assurance program is also proposed for any large scale composting facility.

## ADVANTAGES AND DISADVANTAGES

### Mitigation

In order to review the advantages and disadvantages of mitigation we need to distinguish between the various types of mitigation discussed. In summary, we may divide the different types as follows:

1. Regulatory mitigation.

2. Special purpose mitigation.

3. Performance agreements and penalties.

Regulatory mitigation is more or less prescribed by regulation. Since the regulations for many solid waste management facilities are in a state of flux, however, an implementing agency has a certain amount of discretion in choosing which proposed regulations to follow. As noted earlier, as far as the sanitary landfill is concerned, it is prudent to follow the most stringent proposed regulations.

Special purpose mitigation takes the form of additional visual screening, extending the buffer zone, rerouting traffic, or any number of mitigation measures undertaken to meet a specific concern.

Performance agreements and penalties are mitigation measures proposed in order to meet the public concern that the promises implicit in the proposed mitigation plan may never be kept. The performance agreement states what the promises are, how they will be monitored, how the monitoring data will be shared and the penalties in the form of solid waste management services to be paid if the promises are not met.

The advantages of regulatory mitigation are too numerous to mention. It is sufficient to note that approval at every level depends upon a sensible and thorough application of the regulatory mitigation measures.

The same can be said for special purpose mitigation. Although this level of mitigation is by definition beyond the regulatory requirements, it is vital to the approval process that special concerns be addressed and effectively mitigated wherever reasonably possible.

The principal advantage of performance agreements is that the design and the operating plan for the facility do not specifically identify the levels of protection which the agency undertakes to provide. It is proposed that the performance agreements will be specific. For example, in the case of increased traffic a performance agreement should specify the number of solid waste vehicles that will use a given route during a weekday and during the weekend. Clearly, a range of numbers is necessary, but it should be reasonable. Times of usage should also be identified. Where possible, periods during which there will be no solid waste vehicles should also be specified.

The advantage of including a penalty provision is to provide assurance that the agreement has some effective enforcement, and that if the agreement is broken, there is a clear incentive to correct the problem.

The principal disadvantage of following the most stringent regulatory design and operating requirements is that the cost of operation will be high. In the private sector this would be considered a double-edged sword since on the one hand the most stringent standards provide the greatest protection even including a certain amount of overkill, while on the other hand the costs may be so high that the facility would not be competitive. In the public sector the agency can exercise flow control in order to guarantee that the facility will receive the required tonnage to operate efficiently.

Special purpose mitigation also adds to the overall costs of the facility.

Performance agreements and penalties not only add to the costs of the facility, but they also add a measure of uncertainty for the facility owner in terms of budgeting requirements. To put it simply, there is a risk that a determined group of disgruntled citizens could manipulate the performance feedback system in such a way that the agency would be required to supply free collection and disposal services for an indefinite period.

Compensation

Compensation is also based on a performance agreement. In the case of property value, the performance agreement states there will be no impact and the compensation clause says that if there is impact the party involved will be fully compensated.

In the case of groundwater quality assurance, the performance agreement is even more explicit. The parameters to be monitored and the levels above which pollution is proven are specifically made a part of the agreement.

The advantages of compensation are entirely confined to furthering the ability to site a facility. Most states have significantly impacted the siting process by encouraging considerable public involvment. This has lead to vocal and organized expressions of concern with respect to the risks involved both to the host community and the public citizen. Based upon this experience, it is apparent that expressions of faith in the design and the regulatory process are not of themselves capable of meeting the needs of the concerned citizen. The willingness of the implementing agency to go the extra step and provide compensation in the event of real loss should be a significant factor in reducing public concern.

# Appendix A

| ORDER | U.S. DEPARTMENT OF TRANSPORTATION<br>FEDERAL AVIATION ADMINISTRATION | 5200.5A |
|---|---|---|

1/31/90

1. **PURPOSE.** This order provides guidance concerning the establishment, elimination or monitoring of landfills, open dumps, waste disposal sites or similarly titled facilities on or in the vicinity of airports.

2. **DISTRIBUTION.** This order is distributed to the division level in the Offices of Airport Planning and Programming, Airport Safety and Standards, Air Traffic Evaluations and Analysis, Aviation Safety Oversight, Air Traffic Operations Service, and Flight Standards Service; to the division level in the regional Airports, Air Traffic, and Flight Standards Divisions; to the director level at the Aeronautical Center and the FAA Technical Center; and a limited distribution to all Airport District Offices, Flight Standards Field Offices, and Air Traffic Facilities.

3. **CANCELLATION.** Order 5200.5, FAA Guidance Concerning Sanitary Landfills On Or Near Airports, dated October 16, 1974, is canceled.

4. **BACKGROUND.** Landfills, garbage dumps, sewer or fish waste outfalls and other similarly licensed or titled facilities used for operations to process, bury, store or otherwise dispose of waste, trash and refuse will attract rodents and birds. Where the dump is ignited and produces smoke, an additional attractant is created. All of the above are undesirable and potential hazards to aviation since they erode the safety of the airport environment. The FAA neither approves nor disapproves locations of the facilities above. Such action is the responsibility of the Environmental Protection Agency and/or the appropriate state and local agencies. The role of the FAA is to ensure that airport owners and operators meet their contractual obligations to the United States government regarding compatible land uses in the vicinity of the airport. While the chance of an unforeseeable, random bird strike in flight will always exist, it is nevertheless possible to define conditions within fairly narrow limits where the risk is increased. Those high-risk conditions exist in the approach and departure patterns and landing areas on and in the vicinity of airports. The number of bird strikes reported on aircraft is a matter of continuing concern to the FAA and to airport management. Various observations support the conclusion that waste disposal sites are artificial attractants to birds. Accordingly, disposal sites located in the vicinity of an airport are potentially incompatible with safe flight operations. Those sites that are not compatible need to be eliminated. Airport owners need guidance in making those decisions and the FAA must be in a position to assist. Some airports are not under the jurisdiction of the community or local governing body having control of land usage in the vicinity of the airport. In these cases, the airport owner should use its resources and exert its best efforts to close or control waste disposal operations within the general vicinity of the airport.

5. **EXPLANATION OF CHANGES.** The following list outlines the major changes to Order 5200.5:

   a. Recent developments and new techniques of waste disposal warranted updating and clarification of what constitutes a sanitary landfill. This listing of new titles for waste disposal were outlined in paragraph 4.

   b. Due to a reorganization which placed the Animal Damage Control branch of the U.S. Department of Interior Fish and Wildlife Service under the jurisdiction of the U.S. Department of Agriculture, an address addition was necessary.

   c. A zone of notification was added to the criteria which should provide the appropriate FAA Airports office an opportunity to comment on the proposed disposal site during the selection process.

6. **ACTION.**

   **a.** Waste disposal sites located or proposed to be located within the areas established for an airport by the guidelines set forth in paragraph 7a, b, and c of this order should not be allowed to operate. If a waste disposal site is incompatible with an airport in accordance with guidelines of paragraph 7 and cannot be closed within a reasonable time, it should be operated in accordance with the criteria and instructions issued by Federal agencies such as the Environmental Protection Agency and the Department of Health and Human Services, and other such regulatory bodies that may have applicable requirements. The appropriate FAA airports office should advise airport owners, operators and waste disposal proponents against locating, permitting or concurring in the location of a landfill or similar facility on or in the vicinity of airports.

   (1) Additionally, any operator proposing a new or expanded waste disposal site within 5 miles of a runway end should notify the airport and the appropriate FAA Airports office so as to provide an opportunity to review and comment on the site in accordance with guidance contained in this order. FAA field offices may wish to contact the appropriate State director of the United States Department of Agriculture to assist in this review. Also, any Air Traffic control tower manager or Flight Standards District Office manager and their staffs that become aware of a proposal to develop or expand a disposal site should notify the appropriate FAA Airports office.

   **b.** The operation of a disposal site located beyond the areas described in paragraph 7 must be properly supervised to insure compatibility with the airport.

   **c.** If at any time the disposal site, by virtue of its location or operation, presents a potential hazard to aircraft operations, the owner should take action to correct the situation or terminate operation of the facility. If the owner of the airport also owns or controls the disposal facility and is subject to Federal obligations to protect compatibility of land uses around the airport, failure to take corrective action could place the airport owner in noncompliance with its commitments to the Federal government. The appropriate FAA office should immediately evaluate the situation to determine compliance with federal agreements and take such action as may be warranted under the guidelines as prescribed in Order 5190.6, Airports Compliance Requirements, current edition.

   (1) Airport owners should be encouraged to make periodic inspections of current operations of existing disposal sites near a federally obligated airport where potential bird hazard problems have been reported.

   **d.** This order is not intended to resolve all related problems, but is specifically directed toward eliminating waste disposal sites, landfills and similarly titled facilities in the proximity of airports, thus providing a safer environment for aircraft operations.

   **e.** At airports certificated under Federal Aviation Regulations Part 139, the airport certification manual/specifications should require disposal site inspections at appropriate intervals for those operations meeting the criteria of paragraph 7 that cannot be closed. These inspections are necessary to assure that bird populations are not increasing and that appropriate control procedures are being established and followed. The appropriate FAA Airports offices should develop working relationships with state aviation agencies and state agencies that have authority over waste disposal and landfills to stay abreast of proposed developments and expansions and apprise them of the hazards to aviation that these sites present.

   **f.** When proposing a disposal site, operators should make their plans available to the appropriate state regulatory agencies. Many states have criteria concerning siting requirements specific to their jurisdictions.

**g.** Additional information on waste disposal, bird hazard and related problems may be obtained from the following agencies:

> U.S. Department of Interior Fish and Wildlife Service
> 18th and C Streets, NW
> Washington, DC 20240

> U.S. Department of Agriculture
> Animal Plant Health Inspection Service
> P.O. Box 96464
> Animal Damage Control Program
> Room 1624 South Agriculture Building
> Washington, DC 20090-6464

> U.S. Environmental Protection Agency
> 401 M Street, SW
> Washington, DC 20406

> U.S. Department of Health and Human Services
> 200 Independence Avenue, SW
> Washington, DC 20201

**7.   CRITERIA.** Disposal sites will be considered as incompatible if located within areas established for the airport through the application of the following criteria:

**a.**  Waste disposal sites located within 10,000 feet of any runway end used or planned to be used by turbine powered aircraft.

**b.**  Waste disposal sites located within 5,000 feet of any runway end used only by piston powered aircraft.

**c.**  Any waste disposal site located within a 5 mile radius of a runway end that attracts or sustains hazardous bird movements from feeding, water or roosting areas into, or across the runways and/or approach and departure patterns of aircraft.

*Leonard E. Mudd*

Leonard E. Mudd
Director, Office of Airport Safety and Standards

# References

Algermission, S. T. and D. M. Perkins. A Probabilistic Estimate of Maximum Acceleration in Rock in the Contiguous United States. U. S. Geological Survey Open-File Report 76-416, 1976.

Algermission, S. T. and D. M. Perkins. Probabilistic Estimates of Maximum Acceleration and Velocity in Rock in the Contiguous United States. U. S. Geological Survey Open-File Report 82-1033, 1982.

Aller, L., Bennett, T., Lehr, J. H., and Petty, R. J. DRASTIC: A Standardized System for Evaluating Ground Water Pollution Potential Using Hydrogeologic Settings. National Water Well Association EPA/600/2-85/018, May 1985.

Blokpoel, H. Bird Hazards to Aircraft. Books Canada, Inc., Buffalo, NY 1976.

Bolt, B. A. Earthquakes: A Primer. W. H. Freemen and Company, San Francisco, CA 1978.

Bolt, B. A. et al. Geologic Hazards. Springer-Verlag. Berlin, Germany. 1975.

California Division of Mines and Geology. Guidelines for Evaluating the Hazard of Surface Fault Rupture. CDMG Note Number 49. State of California, Department of Conservation. Sacramento, CA. 1975.

Forsythe. D. M. Gulls, Solid Waste Disposal, and the Bird-Aircraft Strike Hazard. In Proceedings: A Conference On the Biological Aspects of the Bird/Aircraft Collision Problem. Clemson, SC February 5-7, 1974.

Freeze, R. A. and J. S. Cherry. Groundwater. Prentice-Hall, Inc. Englewood Cliffs, NJ, 1979.

Legget, R. F. and P. F. Karrow. Handbook of Geology in Civil Engineering. McGraw-Hill, Inc. New York, N. Y. 1983.

National Oceanic and Atmospheric Administration. Earthquake History of the United States. Publication 41-1. Washington, D. C. 1973

Noble, G. Economic Impact of Prohibiting Landfill Development Within 2000 Feet of Public Schools. Illinois Institute of Natural Resources. 1979.

Solman, V.E.F. Influence of Garbage Dumps Near Airports on the Bird Hazard to Aircraft. Presented at the National Conference on Urban Engineering Terrain Problems. Montreal, Quebec. May 7, 1973.

Terzaghi, K. and R. B. Peck. Soil Mechanics in Engineering Practice. John Wiley and Sons, Inc. New York, NY. 1967

Ultrasystems, Inc. An analysis of Avian Use of the Miliken Sanitary Landfill. Irvine, CA. 1977.

U. S. Department of Transportation, Federal Aviation Administration. Aircraft Bird Strikes Summary and Analysis. Calendar Year 1978.

U. S. Department of Transportation, Federal Aviation Administration. Order 5200.5 FAA Guidance Concerning Sanitary Landfills On or Near Airports. October 16, 1974

U. S. Department of Transportation, Federal Aviation Administration. Order 5200.5A, Waste Disposal Sites On Or Near Airports. January 31, 1990.

U. S. Environmental Protection Agency. Classifying Solid Waste Disposal Facilities (SW-828). Washington, D. C. March 1980.

U. S. D. A. Report ESCS-03. Attitudes of Nearby Residents Towards Establishing Sanitary Landfills. January 1978

U. S. Geological Survey. Facing Geologic and Hydrologic Hazards. Geologic Survey Profession Paper 1240-B. Washington, D. C. 1977.

Wild Scenic Rivers Act. 16 USC Sections 1271 et seq.

# Index

airports                                                        23, 29, 31, 33, 51, 54, 137, 138
anionic exchange capacity                                                                    198
aquifer                                                            4, 6, 10, 11, 13, 15, 39, 43, 45, 199
areas of architectural importance     23, 29, 32, 71, 84, 87, 149, 158, 160, 161, 165, 168, 169, 173, 175,
                                                                           176, 180, 182, 183
artesian well                                                                                 39
ASCE                                                                                          74
availability          3, 24, 30, 32, 38, 41, 42, 71, 77, 86, 92, 95, 96, 98-101, 105, 115, 132, 133,
                                                                           149, 165, 168, 169
bearing strength                                                                          44, 49
bentonite                                                                                     47
bird strikes                                                                              51, 54
branch fault zone                                                                             43
buffer zone                                                   32, 71, 82, 100, 103, 111, 116, 187, 202
building public trust                                                                        187
California Division of Mines and Geology                                              42, 45, 211
compaction densities,                                                                         62
compaction methods                                                                            62
composite map                                                                       132, 137, 139
compost facilities                                                                  2, 145, 161, 168
computer mapping techniques                                                                   57
contingency plan                                                                        196, 197
contractual agreements                                                                   80, 188
cost         14, 32, 40, 56, 57, 63, 77, 80, 84, 86, 101, 105, 106, 115-117, 121, 189, 190, 193,
                                                                           194, 197, 201,202
cover         19, 23, 29, 32, 41, 63, 67, 71, 72, 74, 75, 77, 78, 100, 101, 106, 116,
                                                                           145, 188, 198
cover material                                        23, 29, 41, 72, 74, 75, 77, 101, 106, 116
development                                                   3, 13, 23, 30, 33, 40, 49-51,
     56, 82-84, 95-98, 100-104, 116, 132, 134-136, 138, 139, 149, 153-155, 158, 161, 165, 168, 169, 172, 173,
                                                             175, 176, 180, 182, 183, 185, 192, 193, 200
digitizer                                                                                  15,16
DRASTIC                                                                               3,4,9-12,99
duration limits                                                                              188
Ebenezer Howard                                                                                3
effective porosity                                                                            39
exclusion zones                                    14, 18, 33, 37, 41, 49, 50, 57, 80, 81, 132-134, 137
existing depressions                               23, 29, 32, 71, 77, 95, 100, 101, 104, 107, 116, 149
existing use                                                                               39-41

expansive soils                                                23, 29, 31, 33, 46-48, 116
fault zones                                              12, 23, 29, 31, 33, 42, 43, 116

Federal Aviation Administration                                                      51
Federal Clean Water Act                                                             34
field verification                                                               95, 99
flexible membrane liner                                                         17, 192
floodplains                                                                        132
flow failures                                                                      44
Francis Law Olmsted                                                                 3
frustum                                                                            64
gathering areas                                                            96, 97, 114
groundwater             3, 4, 18, 20, 22, 23, 29, 31-33, 35, 38-41, 47, 55, 71, 80-82, 100, 102-104,
                   109, 110, 111, 116, 137, 146, 149, 162, 164, 165, 168-170, 188, 190, 194, 196,
                                                                             197, 199, 203
groundwater monitoring                                                80, 190, 194, 197
groundwater quality                                                 39, 40, 80, 199, 203
highway restrictions           24, 30, 32, 71, 84, 86, 95, 98, 100, 104, 105, 113, 116, 149, 158,
                                    160, 161, 165, 168, 169, 173, 175, 176, 180, 182, 183
holocene                                                                        12, 43
host community fee                                                             192-196
hydraulic conductivity                                                      4, 9-11, 39
hydraulic gradient                                                                 39
hydric soils                                                                       35
hydrophytic vegetation                                                             35
Ian McHarg                                                                         3
Illinois Archaeological Survey                                                    134
Illinois Department of Conservation                                               133
Illinois Department of Energy and Natural Resources                          131, 137
Illinois Pollution Control Board                                              134, 135
Illinois State Geological Survey                                                   133
Illinois State Natural History Survey                                              133
in-place volume                                                                63, 64
incinerators                                                                2, 145, 154
intrinsic Suitability                                                              12
inward gradient design                                                            82
jet aircraft                                                                   51, 137
Joint Action Solid Waste Planning Agency                                          131
karst                                                                      6, 8, 13, 46, 48
landslide prone areas                                                          46, 49
lateral spreads                                                                    44
leachate          1, 19, 22, 36, 40-43, 45, 63, 64, 80, 82, 103, 109, 188, 190, 193, 194, 197, 198

| | |
|---|---|
| liner integrity | 191 |
| liquification | 44 |
| local study | 14, 18, 89, 92 |
| loess | 47, 49 |
| main fault zone | 43 |
| Minnesota Pollution Control Agency (MPCA) | 12 |
| mitigation | 14, 18, 101, 117, 192, 194, 195, 197, 201-203 |
| montmorillonites | 47 |
| municipal wells | 31, 33, 41, 55, 116, 137, 138 |
| National Historic Preservation Act | 83 |
| natural screening | 23, 29, 32, 71, 77, 78, 95-97, 100, 101, 103, 107, 116, 149, 153, 154, 158, 160, 161, 165, 168, 173, 175, 176, 180, 182, 183 |
| New York State Department of Health | 74 |
| Pennsylvania Department of Environmental Resources | 74 |
| performance agreements | 202, 203 |
| permeability | 17, 18, 39, 40, 42, 55, 77, 82, 103, 133 |
| pleistocene | 43 |
| pollution potential | 3, 4, 9 |
| prime farmland | 23, 29, 31, 33, 55, 56, 153, 154 |
| problem definition meetings | 120 |
| progress reports | 191 |
| propeller driven aircraft | 51 |
| proximity to major highway | 24, 30, 31, 33, 56, 59, 136, 138, 139, 149, 160, 165, 168, 169, 175, 182 |
| public meetings | 123, 129, 187 |
| public participation | 119, 126, 128, 130 |
| quality assurance programs | 188, 191 |
| rating | 4-11, 99, 100, 102-116, 122, 126-130 |
| recycle factor | 62 |
| recycling facilities | 2, 145, 176 |
| regional criteria | 31-33, 36, 39, 57, 81, 103, 131, 132, 153, 168, 169 |
| regional siting map | 60, 87, 122 |
| regional study | 14, 18 |
| residential well density | 23, 29, 32, 71, 80, 94, 100, 102, 108 |
| run-on potential | 32, 71, 78, 80, 81, 97, 100, 102, 108, 116 |
| scenic area | 81, 103, 110, 153, 154, 169 |
| screening plan | 78 |
| secondary fault zone | 43 |
| seismic impact zones | 23, 29, 31, 33, 43 |
| seive map | 139 |
| sewage treatment facilities | 2, 145 |

# About the Author

George Noble is a registered professional engineer, and a diplomate of the American Academy of Environmental Engineers.  He lives in Wilmette, Illinois with his wife Paula and two children, Tracy and David.